Colour Vision
in the
Nineteenth Century

Colour Vision in the Nineteenth Century

THE YOUNG–HELMHOLTZ–MAXWELL THEORY

Paul D Sherman

Foreword by W D Wright

Adam Hilger Ltd, Bristol

British Library Cataloguing in Publication Data

Sherman, Paul D.
Colour vision in the nineteenth century.
1. Colour vision
I. Title
612′.84 QP483

ISBN 0-85274-376-9

Published by Adam Hilger Ltd,
Techno House, Redcliffe Way, Bristol BS1 6NX

The Adam Hilger book-publishing imprint is owned by
The Institute of Physics

Printed in Great Britain by Pitman Press Ltd, Bath

To my parents,

Riva and Leo Sherman,

who were the ultimate source of inspiration for this book

Contents

Foreword

If other readers enjoy this book as much as I have, it will be very popular indeed. The story that Paul Sherman has to tell is fascinating and in my view he has done a great service in giving us a new and truer perspective of the Young–Helmholtz–Maxwell school of colour perception. (The schools of other nineteenth-century giants, notably of Goethe and of Herring, are not dealt with here.) Maxwell's massive contributions to theoretical physics and optics have tended to overshadow the very important contributions which he also made to the science of colour, and I am sure we have sometimes attributed to Helmholtz concepts and ideas which were really due to Maxwell. Grassman, too, is another figure to whom we pay lip service without knowing too much about him and, with this book in our hands, we can give credit where credit is due. As Sherman's narrative unfolds, we are able to follow the trails laid by Brewster, Wollaston and other less familiar figures. Scientific knowledge does not progress steadily forwards and upwards, and this historical study provides sobering examples of false tracks and fruitless arguments. There are lessons here for all of us to learn.

W D Wright

Preface

The nature of colour vision has always been an absorbing interest for scientist and non-scientist alike. From Aristotle to Schrödinger, physicists and physiologists, artists and philosophers have all written about it so that the subject transcends all traditional boundaries and in its eclectic character lies its fascination and its deep challenge. Yet it is remarkable that much of the history of colour vision actually falls within the scope of physics, since optical measurements provide considerable knowledge about colour vision processes. It is this history, from about 1800 to 1860, which I have chosen to relate, especially as it pertains to the creation of the modern subject. In this period three immortals of physics, Thomas Young, Hermann von Helmholtz and James Clerk Maxwell, brought the infant science of colour perception to a new modern understanding.

The story opens with a famous speculation by Young in 1801 on the physiology of colour vision in terms of three primary colours, which in itself was a curious notion. That there were indeed three primary colours was an accepted impression in Young's time but it was more folklore than science. It remained for the German mathematician-philologer Herman Günther Grassmann over fifty years later to demonstrate the fact of three-fold vision and then from a quite unexpected approach. Still, Young saw in this sentiment the germ of a theory of colour vision. Presented in the context of his undulatory theory of light, Young's trichromatic theory of vision had just enough plausibility to stimulate continuing discussions. In Britain and elsewhere Young's theory was always acknowledged and referred to whenever questions on vision arose. Moreover, it influenced David Brewster in his researches on the solar spectrum and it played a role in Brewster's famous three-colour spectrum. Partly in response to Young's theory (which he rejected) Brewster evolved his own ideas of colour perception based on his supposed observation of the solar spectrum in three separate overlapping spectra of red, yellow and blue. That idea raised considerable controversy among physicists, some of whom supported Brewster but most of whom viewed his theory as a singular error. None of them realised, however, that their arguments over Brewster's theory really concerned the very nature of colour sensation itself, complicated by an erroneous assumption about colour mixtures. In that respect a parallel line of enquiry studied colour combinations and colour classifications, occupying such notable persons as Tobias Mayer and Johannes Lambert in the eighteenth century and John Herschel and M E Chevreul in the early nineteenth century. They too, alas, were subject to the false assumption that pervaded debate on Brewster's theory, but at least in those researches Brewster's theory was not the obstacle that it became in understanding vision as such.

A peculiar difficulty underlay all these problems in that no reliable repeatable experimental methods existed either for compounding colour combinations or for determining the sensitivity of the eye to individual colours. Without an experimental basis, hardly any theoretical understanding was possible. Hence the study of colour combinations and colour perception was perforce a descriptive science concerned more with qualitative observations and speculative hypotheses than with exact measurements and mathematical expression. This being the case it is remarkable that it should have attracted the mathematical physicists; yet it did precisely that.

The fog of controversy over Brewster's theory lifted in 1852 when Helmholtz showed that Brewster had been victim of several subjective illusions concerning colour and that the spectral colours were all still in their rightful places. He showed furthermore that the disputants had always been equating two quite different physical processes, additive and subtractive colour mixing (see Plate 1), and were thereby adding to their confusion. His distinction of these processes marked a turning point in understanding colour mixtures and hence colour vision.

Ultimately, however, it was James Clerk Maxwell who demonstrated how to measure the subjective colour sensations. In a series of important investigations Maxwell showed how to use known optical methods to measure the eye's response to colour. He confirmed Young's early speculation about trichromatic vision together with his choice of primary colours, and he showed that colourblindness, a much discussed subsidiary problem, could be understood in terms of a missing sensation, as Young had predicted. Far more than Helmholtz, it was Maxwell who revived Young's hypothesis—not that it had ever really needed reviving—and made it a valid basis for further work. Maxwell's experimental methods did nothing less than transform a qualitative science into a quantitative one with a whole new range of experimental techniques.

Maxwell is famous, of course, as the founder of the electromagnetic and kinetic molecular theories. In exactly the same sense he is a founder of a modern active field of research—that of colour vision and colour specification, which is not, incidentally, simply a matter of measuring wavelengths. It is Maxwell's methods of colorimetry which underlie many of today's methods of specifying colour. The modern technologies of colour films and colour television, for example, are hardly imaginable without the ability to define colours as first demonstrated by Maxwell. Maxwell contributed to essentially every area of physics in his day and the physics of colour vision was one of them. For him, colour was a continuing, indeed life-long, interest representing an aspect of his career which is hardly known today. In short then, the events described in this book form the prelude to modern research in colour perception.

In preparing this study I am deeply indebted to many people for their kindness and generosity. Professor A Rupert Hall read an early version of the manuscript and gave most generously and graciously of his time and advice in improving it. Professor W D Wright read the final version and provided guidance in technical matters where I would otherwise have been in error. Needless to say, of course, any other errors are mine alone. Mr Leslie Jones and Mr Gerald Fishman drafted many of the figures and Mr Lawrence Byrwa unstintingly helped in preparing numerous other illustrations. Dr Berne Dibner of the Burndy Library provided me with office space and placed his resources at my disposal while Mrs Leni Greenwood of the Burndy Library gave much help with German translation. Mr Chip Chillington of the Library was also a source of encouragement. Mrs Paula Tarantino very helpfully typed part of the manuscript when I was struggling to meet a deadline. Last but not least I must acknowledge my wife for her patience and understanding. To all these people I am most grateful and extend my sincerest thanks.

Paul D Sherman
Eastchester, NY

1
Thomas Young: His Speculation About Three-colour Vision

In the year 1772 the mechanism of vision was almost completely understood. No less an authority than Joseph Priestley, the discoverer of oxygen and author of a definitive survey on light and vision wrote that 'the nature of vision, in general, seems to be very well understood; and there are few phenomena belonging to it that have not been satisfactorily explained . . .' [1] If the eye were considered as a camera obscura, its image forming properties could be reduced to well known principles of geometrical optics. Indeed the essence of Priestley's discussion of the eye's optical properties is still found in elementary physics textbooks so, for the modern physics student, knowledge of this subject initially extends only to the late eighteenth century.

For Priestley there were only two outstanding questions concerning vision, and each had two possible answers. The first question involved the means by which the eye accommodates its focus for near and far objects. One possible solution considered that the crystalline lens changed its focal length to accommodate a fixed lens–screen distance; the second considered that the eyeball itself changed length thereby altering the lens–screen distance for a fixed-focus lens. The other unanswered question involved the nature of the image-receiving screen itself. What part of the eye actually received the image, the retina or the choroid? The evidence even then suggested the retina as the answer, but that remained to be seen. Priestley's view in any case was that, although neither question was completely solved, the essence of solution for each was in hand. Therefore he could say with equanimity that there were few phenomena involving vision which had not been satisfactorily explained. No question about vision in colour was asked or referred to, indicating that it was not then an issue.

Since the question of colour vision was hardly even recognised in 1772, naturally it was yet to be a subject of real investigation. However, a few people had thought of the possibility. Isaac Newton, for example, had speculated in 1704 in his *Opticks* that colour vision resulted from vibrations of some sort excited 'on the bottom of the eye' by the arriving light. [2, Query

Thomas Young
(courtesy of Burndy Library).

no 12, p345] but he was necessarily vague, not having really investigated this question nor having the means to do so. His proposal was a speculation only. Forty-five years later David Hartley, the philosopher, published a work entitled *Observations on Man* in which he attempted to carry Newton's idea further. He considered that the distinction of colours is specified by their ratios, which in turn could be determined from the proportions of their separate sections in the spectrum as measured by Newton. Hartley supposed that light stimulated vibrations in the retina with the ratio between red and violet vibrations being 1:2. [3] He did not consider making the necessary observations himself, being content to let Newton's observations stand. Nor did he progress beyond this limited speculation; his main interest was ethics and this hypothesis was only part of a general one by which he explained all mental phenomena in terms of minute vibrations. Priestley himself was sympathetic to Hartley's idea but, as a productive scientist rather than a philosopher, he was more cautious of such universal nostrums as workable ideas. Friend to Hartley though he was, Priestley did not include this speculation in his survey of optics.

Fourteen years after Priestley's writing, the subject of colour in relation to vision was raised ever so briefly by Robert Waring Darwin (father of the great naturalist, Charles). Describing a series of ocular experiments, Darwin concluded that the retina is responsible for the sensation of vision. [4, p314] But there were no 'minute nervous vibrations' for Darwin. He meant that the retina was literally stimulated into motion by the action of light, either vibrating in some unexplained manner or changing its thickness at different places. Therefore an image on the retina could result in its taking on different thicknesses at different points along the image. Such thinking led Darwin to an analogy with Newton's experiments on thin films. Since different colours are reflected or transmitted according to the different thicknesses of the film as Newton had shown, Darwin said 'is it not probable, that the effect of the retina may be to alter its thickness or thinness, so as better to adapt it to reflect or transmit the colours which stimulate it into action?' [4, pp347–8] In the end it was really not a hypothesis of colour vision for Darwin but an adaptation of Newton's theory of fits of easy reflection and transmission to the process of vision. It remained for Thomas Young in 1801 to create the modern subject of colour perception. And then he did so in true eighteenth century fashion, invoking vibrations in the retina while also invoking Newton's ghost.

Young was certainly one of the acutest men who ever lived. Born in 1773 he was by the age of twenty master of almost every field of knowledge. By turns he made significant contributions to medicine, philology, physics, actuarial calculations, mathematics and Egyptology. He reminds one of Eratosthenes, the famed librarian of Alexandria in the third century BC, who also contributed distinctively to many fields. Eratosthenes was known by the nicknames of Pentathlos and Beta because of his many sided interests and because he was never the absolute leader in any one field. It has been said of Young by another 'pentathlos' (Hermann von Helmholtz) that, had he concentrated on only one or two of his many areas, he must certainly have been the foremost leader of each. For example, as a physician and physicist he solved the outstanding question of visual accommodation, confirming Priestley's confident expectations; as a philologist Young was the first to show in 1814 how to read Egyptian hieroglyphics but he published very little about it; as a physicist he is known for Young's modulus of elasticity and, more importantly, his name is forever linked with the wave theory of light. It was he who conceived the idea of light wave interference and introduced it to physics in a series of famous papers. It was within the context of the undulatory theory of light, for so the wave theory was then called, that Young initiated the modern theory of colour perception.

The eye and the mechanism of vision had fascinated Young from his earliest medical education. Entering St Bartholomew's Hospital, London, in September 1793 as a student, he undertook the usual regimen of anatomy, physiology and 'midwifery'. A regular procedure of the anatomy course included dissecting an eye (not necessarily human) and here Young had a marvellous piece of luck. Upon receiving the eye of a freshly killed ox he found that the ciliary muscles attached to the lens were still active. He noted that they could change the shape of the lens, thus changing its focal length and this, of course, Young recognised accounted for accommodation. While still a medical student he embarked on a detailed study of the subject, the result of which was a long paper delivered to the Royal Society in 1793. Young demonstrated there the correct solution to the question of accommodation together with an analysis and refutation of alternative speculations by earlier investigators. So highly regarded was this work that he was elected a Fellow of the Royal Society at the age of 21.

But Young was also a physicist. While completing his formal medical studies at Cambridge in the years 1797–9, he found time for extensive experiments on sound and light, the results of which formed the basis of a long paper on acoustics in 1800, again presented to the Royal Society, [5, p106] and here the germ of his thinking about light in terms of waves began to appear. In a disciplined, informed manner, Young drew a strong analogy between sound as an undulation in air and light as a possible undulation in an ether medium. Although it was not original to him, nor did he present any really new evidence for it, his analogy was typically closely and strongly argued. Ultimately it served to direct his attention to a more thorough study of light in terms of undulations or waves.

In the same period, Young returned to the subject of the eye. Indeed, in these years Young was deeply occupied by both physics in terms of light and physiology in terms of vision. In 1801 he published a monumental paper entitled *On the Mechanism of the Eye*. [6] He answered unequivocally some objections to his earlier work on accommodation, so there could no longer be any doubt of the action of the ciliary muscles and the change of shape of the lens in the process. Beyond this the paper discussed essentially every known aspect of vision, both optical and anatomical, and is particularly noteworthy for the first clinical description of astigmatism (in his own eye), a visual problem previously unrecognised. Not surprisingly, considering all these studies, Young is often considered the father of physiological optics.

Young's interests (light, vision and sound) finally merged in another epoch-making paper scarcely a year later. Delivering his second Bakerian Lecture to the Royal Society in 1801, Young proposed his celebrated undulatory (or wave) theory of light which is now so much associated with his name. [7] It was however, as Young himself said, not an entirely new theory but more intended to 'refer some theories . . . to their original inventors, support them by additional evidence and to apply them to a great number of diversified facts which have hitherto been buried in obscurity.' [7, p12] It was in this context that Young proposed both his wave theory of light and his theory of three-colour vision. Certainly the new theory of vision, if not that of waves, lay buried in obscurity, being essentially an untouched idea.

After a few introductory remarks, Young initiated his undulatory theory by first hypothesising a universally pervading luminiferous ether and next hypothesising that 'undulations are excited in the ether whenever a Body becomes luminous.' [7, p16] Support for both these hypotheses consisted entirely of long quotations from Newton, taken out of context and rearranged to suit Young's ends, thereby making Newton appear (as was Young's plan) to be the original inventor of the theory. These two hypotheses were hardly new, especially in view of the excerpts quoted by Young. Indeed they had been seen several times during the seventeenth and eighteenth centuries, but now there was a subtle difference. For the earlier physicists, except possibly Christian Huygens, such statements regarding an ether medium for light were little more than speculation without very serious content. For Newton especially, an ether played a different role in the propagation of light than in a wave theory itself. For Young, however, the ether and its undulations were a physical reality. In fact, in his 1800 paper on sound and light he had written 'that a medium resembling, in many properties, that which has been denominated ether, does really exist, is undeniably proved by the phenomena of electricity.' [5, p126] Here at least an ether was generally regarded as necessary in transmitting electric effects. Yet even so the electric ether was at best a vague concept discussed in qualitative terms only. Nevertheless, most physicists thought of it as a reality and Young in making this comparison is part of the eighteenth century tradition. In any case, he tactily assumed that the ether presumed to exist for electric effects must be the same ether for light and this was certainly a new conception. Whether or not this were true, Young certainly started a new fashion, for all his nineteenth century successors also regarded the ether as a basic physical reality. Like Young, they could only visualise waves of light as vibrations of an elastic medium.

Having established the existence of undulations in an all-pervading ether as the basis for a theory of light, Young immediately proceeded to an application of the theory according to his stated plan. In his third hypothesis, he gave the essence of his theory of colour vision:

Isaac Newton
(courtesy of Burndy Library).

> The Sensation of different Colours depends on the different frequencies of Vibration, excited by Light in the Retina. [7, p126]

Continuing his programme of referring his theories to their original inventors, Young quoted yet another long series of passages from Newton again out of context and rearranged. For example, immediately following the hypothesis above, Young quoted Newton as follows:

> The objector's hypothesis, as to the fundamental part of it, is not against me. That fundamental supposition is, that the parts of bodies when briskly agitated, do excite vibrations in the ether, which are propagated every way from those bodies in straight lines, and cause a sensation of light by beating and dashing against the bottom of the eye, something after the manner that vibrations in the air cause a sensation of sound by beating against the organs of hearing. Now the most free and natural application of this hypothesis to the solution of phenomena, I take to be this: that the agitated parts of bodies, according to their several sizes, figures, and motions, do excite vibration in the ether of various depths or bignesses, which, being promiscuously propagated through that medium to our eyes, effect in us a sensation of light of a white colour; but if by any means those of unequal bignesses be separated from one another, the largest beget a sensation of a red colour, the least or shortest of a deep violet, and the intermediate ones of intermediate colours; much after the manner that bodies, according to their several sizes, shapes and motions, excite vibrations in the air of various bignesses, which, according to those bignesses, make several tones in sound . . . [8, p5088]

The original date of this writing was 1672 and it seems superficially that Newton was advocating a wave theory of light. But Newton was actually defending here quite another matter. Robert Hooke had taken Newton to task for supposedly espousing a theory of corporeality of light in his first optical paper. Insisting that his own hypothesis of pulses or waves in a uniform medium was just as feasible an explanation of light as Newton's corpuscles, Hooke had even implied that it was better. Newton then in the passage above was replying to Hooke, defending himself against some rather nasty and unwarranted criticism. Newton was showing that he had only suggested the possibility of light corpuscles and that actually such a possibility had quite some similarity to Hooke's own concept. For just before the passage above Newton wrote:

> For certainly it has a much greater affinity with his own Hypothesis, than he seems to be aware of; the Vibrations of the Aether being as useful and necessary in this, as in his. For, assuming the Rays of Light to be small bodies, emitted every way from Shining substances, those, when they impinge on any Refracting or Reflecting superficies, must as necessarily excite vibrations in the aether, as Stones do in water when thrown into it. [8, p5087]

Then the passage quoted by Young followed (after two more sentences) in order to amplify this thought. In this context, Newton's meaning in the passage quoted by Young is much clearer; but for Young it well suited his purpose, illustrating from the Master the point Young himself wished to make—light consists of vibrations in an ether while vision consists of vibrations in the retina.

The passage from Newton quoted above by Young was followed closely by a second quotation from Newton:

> To explain colours, I suppose, that as bodies of various sizes, densities, or sensations, do by percussion or other action excite sounds of various tones, and consequently vibrations in the air of different bigness; so the rays of light, by impinging on the stiff refracting superficies, excite vibrations in the ether . . . of various bigness; the biggest, strongest or most potent rays, the largest vibrations; and others shorter, according to their bigness, strength, or power; and therefore the ends of the capillamenta of the optic nerve, which pave or face the retina, being such refracting superficies, when the rays impinge upon them, they must there excite these vibrations, which vibrations (like those of sound in a trunk or trumpet) will run along the aqueous pores or crystalline pith of the capillamenta, through the optic nerves, into the sensorium . . . and there, I suppose, affect the sense with various colours, according to their bigness and mixture; the biggest with the strongest colours, reds and yellows; the least with the weakest, blues and violets; the middle with green; and a confusion of all with white, much after the manner that, in the sense of hearing, nature makes use of aerial vibrations of several bignesses, to generate sounds of divers tone; for the analogy of nature is to be observed. [9]

Newton has here implied an analogy between sound and light in terms of waves. He has not, however, so far as Young's quotation would show really explicitly compared waves of sound with possible waves of light. This Newton did in the next few sentences: 'And further, as the harmony and discord of sounds proceed from the proportion of the aereal vibrations, so may the harmony of some colours, as of golden and blue, and discord of others, as of red and blue, proceed from the proportions of the aethereal.' [9] This passage, however, Young omitted contenting himself with implied analogy only. But for Young, that analogy with nature was the *only* thing observed; for ultimately it was the merging of his interests in the eye and in optics, together with his earlier work in acoustics, which led Young to these ideas. But of proof there was none. He could only argue the point.

Young's quotations from Newton above were followed closely by his last passage from the latter's work:

> Considering the lastingness of the motions excited in the bottom of the eye by light, are they not of a vibrating nature?—Do not the most refrangible rays excite the shortest vibrations—the least refrangible the

largest? May not the harmony and discord of colours arise from the proportions of the vibrations propagated through the fibres of the optic nerve into the brain, as the harmony and discord of sounds arise from the proportions of the vibrations of the air? [2]

These remarks are taken from Newton's *Opticks* where they are found separately as 'Querries' number 16, 13 and 14, respectively. Young has taken only part of each Querry and put them together in his own order to make a coherent argument in support of his own ideas. For Newton they were, as the term Querry implies, matters of conjecture only. But for Young, quoting them out of order and out of context, they formed a coherent argument in support if his hypothesis. Interestingly enough, of the three hypotheses introducing his wave theory, all so deeply dependent on Newton for viability, this third one invokes by far the longest series of quotations. With little else to justify his theory, Young felt that this hypothesis more than the others needed additional support.

All this is reminiscent of the Scholastic philosopher appealing to ancient authority for support. Whereas the Scholastics often took Aristotle as their 'Master', Young chose Newton. Of course, he had good reason for doing so, for Newton's reputation at the time was supreme in matters of mechanics and light. The long quotations, of course, were really intended to buttress Young's own views rather than referring them to their original inventor's. Regardless of what Young has made of these passages, it is clear that Newton had stated them merely as matters of argument or supposition; for him they represented no final opinion. Moreover, Newton was viewed by Young's contemporaries as having embraced a purely corpuscular theory of light without reference to an ether or vibrations. Therefore a wave theory not in accord with the 'Master's' thinking as interpreted at that time was certain to be unpopular. Hence Young thought it necessary to show that Newton too had employed undulatory or vibrational ideas of light and in this he was partly correct. However, in order to make his own ideas more palatable he cast Newton as an espouser of a pure wave theory in the sense of Christian Huygens, and this he certainly was not. Even Young had lingering doubts, for he conceded toward the end of the paper that Newton was not truly a proponent of the wave theory in the sense that he was proposing. For Young, the undulations of the ether were the light itself; but although Newton had conceived of vibrations in an ether he denied that they actually constituted the light. In all candidness, Young was forced to admit this fact. [7, p45] So in effect, Young had come full circle; having attempted to show that Newton was an adherent or even creator of an undulatory theory, Newton at the last was necessarily deprived of that honour. The result of it all was that Young's attempt to cast Newton as a proponent of the wave theory of light failed utterly and was even counterproductive. It produced a vituperative denunciation of Young from Henry Brougham from which Young never quite recovered. In the end it was Augustin Fresnel, not Young, who really placed the wave theory of light on a firm theoretical basis. Interestingly enough, however, Young did have one single anticlimatic success. Sir George Biddell Airy, Astronomer Royal and early adherent of the wave theory, wrote in 1833 'I think that Dr Young has fully made out . . . that Newton was a believer in the theory of undulations.' [10] But by then Young was dead and the wave theory was well on its way toward general acceptance owing more to its mathematical formulation by Fresnel than to Young's arguments.

The rest of Young's theory of colour vision takes place in a short scholium to the third hypothesis. Assuming *a priori* that sight itself consists of some sort of motion in the retina, two questions arose: what types of motion are most suitable to the retina, and how shall the retina be constructed to allow these motions to take place? To the first question Young proposed, no doubt with Darwin's paper in mind, that the motion of the retina is 'rather of a vibratory than an undulatory nature.' [7, p20] It is akin to a resonant motion with points on the retina 'vibrating in perfect unison with every possible undulation'. [7, p20] This question settled, one was led directly to that of the retina's construction.

Given the undulatory nature of light, which Young assumed, different coloured lights would have to consist of undulations of different wavelengths. Indeed, he even calculated a table of colours in terms of wavelengths from Newton's measurements on thin films (now called Newton's rings). [7, p39] [11] But as there can be an infinite variety of wavelengths and hence colours, each sensitive point of the retina on first thought would have to be capable of an infinite variety of vibrations. Young found this impossible to imagine and so he proposed an alternative scheme.

It had been known since Newton's early optical papers that different prismatic colours could be superimposed to give intermediate colours. Moreover, Newton had shown that all the spectral colours could be recombined to give white light. Finally, Newton had implied that white light could be made from fewer than the seven spectral colours although the minimum number of colours required he did not know. However, Newton had said that at least three colours were necessary to produce white light, since there was already an established tradition in his time that three primary colours only were needed to reproduce all hues (see Chapter 4). Therefore, Young proposed that each point of the retina be 'put in motion more or less forcibly' by only three 'principal' colours. [7, p21] He chose red, yellow and blue, generally accepted by artistic opinion at that time as the primary colours.

He proposed that if the wavelengths of red, yellow and blue were in the ratio of 8:7:6 (or nearly) as determined from his table of wavelengths, then the eye would see all colours created by such vibrations or by vibrations in ratios close to these. For example he stated:

> the undulations of green light being nearly in the ratio of $6\frac{1}{2}$, will affect equally the particle in unison with yellow and blue, and produce the same effect as a light composed of those two species; and each sensitive filament of the nerve may consist of three portions, one for each principal colour. [7, p21]

Thus Young conceived colour vision as a kind of colour mixing process in the retina. Almost certainly this was based on the continuing analogy between waves of sound and presumed waves of light. Young had discussed at length in his 1800 paper on sound and light the phenomenon of beat frequencies whereby two notes of different frequencies can together produce a third tone whose frequency is the difference between the two. While the retinal colour mixing proposed here is not the same kind of process, for Young, master of both physiological optics and acoustics, the acoustical effect could be modified and moulded to fit the eye. As mentioned before, it was a merging in Young of these interests which originally led to this paper.

THE CHOICE OF PRINCIPAL COLOURS REVISED

About seven months after Young's Bakerian Lecture of 1801, new observations on the structure of the sun's spectrum caused Young to reconsider his choice of principal colours. William Hyde Wollaston had been pursuing a series of researches on the refractive and dispersive powers of prisms, the results of which he read before the Royal Society on June 24, 1802. [12, p365] In a passage relegated to the very end of the paper, Wollaston stated,

> I cannot conclude these observations in dispersion, without remarking that the colours into which a beam of white light is separable by refraction, appear to me to be neither 7, as they are seen in the rainbow, nor reducible by any means (that I can find) to 3, as some persons have conceived; but that, by employing a very narrow pencil of light, 4 primary divisions of the prismatic spectrum may be seen, with a degree of distinctness that, I believe, has not been described nor observed before. [12, p378]

Wollaston was referring to an experiment in which he had admitted sunlight into a darkened room through a slit 1/20 in wide. Some 10 to 12 ft beyond the slit the sunlight was incident on a flint glass prism 'free from veins' which he held in front of his eye. [12, p378] The arrangement is shown in Figure 1.1.

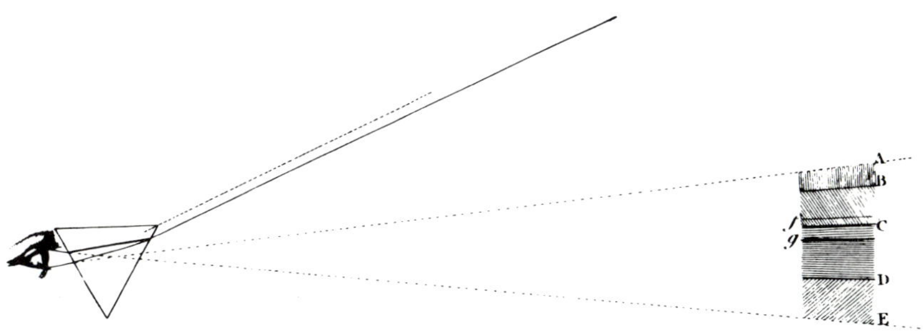

Figure 1.1 Wollaston's drawing of the spectrum showing his partition into four 'primary divisions'.

William Hyde Wollaston
(courtesy of Burndy Library).

Viewed in this manner, Wollaston stated 'the beam is seen to be separated into the four following colours only, red, yellowish-green, blue and violet.' [12, p378] Crossing the spectrum were several dark lines described by Wollaston as follows:

> The line A that bounds the red side of the spectrum is somewhat confused, which seems in part owing to want of power in the eye to converge red light. The line B, between red and green, in a certain position of the prism, is perfectly distinct; so also are D and E, the limits of violet. But C, the limit of green and blue, is not so clearly marked as the rest; and there are also, on each side of this limit, other distinct dark lines, *f* and *g*, either of which, in an imperfect experiment, might be mistaken for the boundary of these colours. [12, p378]

Clearly what Wollaston viewed here were several Fraunhofer lines. But whereas some years later these were correctly interpreted as absorption lines in the sun's atmosphere, Wollaston's interpretation of the lines, and hence the structure of the spectrum, was quite different.

Having attained a degree of dispersion beyond that of Newton or anyone else previously, he thought he had found a more fundamental composition of the spectrum. For Wollaston, the sun's spectrum correctly consisted of four colours, not seven as Newton had shown, and these had to be red, yellowish-green, blue and violet. The dark lines were then dividing lines or boundaries between the different parts of the spectrum. This is clear not only from the remarks made above but also from his consistent usage of terminology indicating the lines as divisions between the colours. He even stated how 'the colours are most clearly divided' in a passage on the experimental techniques he employed. [12, p379] Thus whereas he somewhat pompously warned his listeners that the lines he called *f* and *g* were not boundaries between colours, he himself was a victim of that very error.

For confirming evidence on spectra in general, Wollaston also examined the spectrum of the lower part of a candle flame and the 'blue line of electric light', [12, p380] presumably a spark from a static electric discharge. Describing each case he consistently used terminology indicating a divided, sectioned spectrum so he certainly thought he had discovered something new concerning the structure of the solar spectrum. For him the spectrum was a discontinuous phenomenon as opposed to the contiguous hues seen by his predecessors. Therefore, Wollaston interpreted the four colours he saw in the sun's spectrum as simply its purest colours. There was no 'pure' orange, yellow or indigo in the sun's visible spectrum according to his idea.

Wollaston's paper was followed only seven days later by Young's next major effort on the wave theory of light. [13] Here as in his previous paper (the Bakerian Lecture of 1801), Young's primary aim was to advance his undulatory theory of light rather than discuss colour vision, so the parts of importance to vision were relegated to a somewhat inconspicuous portion of a paper really devoted to another topic. Mentioning Wollaston's 'very interesting observations' [13, p394] Young reported that he had repeated the experiments with 'perfect success'. [13, p394] (Fast work indeed.) Young, believing in the correctness of Wollaston's spectral interpretation, on the basis of this alone modified his choice of principal colours for colour vision. Said Young,

> In consequence of Dr Wollaston's correction of the description of the prismatic spectrum, compared with these observations, it becomes necessary to modify the supposition that I advanced in the last Bakerian lecture, respecting the proportions of the sympathetic fibres of the retina; substituting red, green and violet for red, yellow, and blue, and the numbers 7, 6, and 5 for 8, 7, and 6. [13, p395]

Young's choice of red, green and violet (see Plate 2) for the primary colours is curious. Wollaston after all had mentioned red, yellowish-green, blue and violet as the spectral divisions and seemingly Young had also achieved the same 'perfect' results. A clue to Young's thinking is given in his *Course of Lectures on Natural Philosophy and the Mechanical Arts*, published in 1807. Here again he referred to Wollaston's experiment and his 'more effectual separation of the colours' [14, vol I, p343] than that of Newton or anyone to date. Taking over Wollaston's interpretation completely, Young stated,

> The spectrum formed in this manner consists of four colours only, red, green, blue, and violet, which occupy spaces in the proportion of 16, 23, 36, and 25, respectively . . . the colours differ scarcely at all in quality within their respective limits, but they vary in brightness; the greatest intensity of light being in that part of the green which is nearest to the red. A narrow line of yellow is generally visible at the limit of the red and green, but its breadth scarcely exceeds that of the aperture by which light is admitted, and Dr Wollaston attributes it to the mixture of the red and the green light. [14, vol I, p343]

The remark about the presence of a yellow line in the solar spectrum and its constitution is something new. Describing the solar spectrum, Wollaston had not mentioned such a line although he did refer to a yellow line in his candle spectra, probably due to sodium. However, Young was referring here to Wollaston's solar observation only since the spaces occupied by each colour as he gives them are taken directly from Wollaston's description of the solar spectrum. Furthermore, Wollaston did not originally conceive any of his spectral colours as mixtures of light. Colour mixtures seem to have been Young's speciality. With no correspondence existing between the two men, one can only guess that Wollaston's ammended interpretation was a later thought stimulated by Young himself perhaps and communicated to his friend privately. However Young came to choose red, green and violet as his primary colours, he found it

> certain that the perfect sensations of yellow and of blue are produced respectively, by mixtures of red and green and violet light, and there is reason to suspect that those sensations are always compounded of the separate sensations combined . . . and we may consider white light as

> composed of a mixture of red, green and violet only . . . [14, vol I, p345]

Most important to this interpretation was a series of confirmatory experiments on colour mixing performed by Young. For the theory to work, the three colours chosen had to be combinable into white light as Young implied above. At sometime between 1802 and 1807 Young performed colour mixing experiments by painting circular discs with different colours 'in any way that may be desired.' [14, vol I, p345] and then revolving the discs. A spinning disc painted with sectors of different colours gives a single hue due to the combined effect of the coloured sections. Such a series of experiments can produce the colour combinations mentioned by Young, and with a single disc painted with sectors of red, green and violet in proper proportions one can produce a greyish imitation of white. Such combinations, except the greyish white, were shown by Young in his *Lectures*. [14, vol II, plate xxix and figures 423–6] It is, however, highly doubtful if Young ever produced even a good approximation to white in this manner as he implied he did in describing his experiments. For, as we shall see, even Maxwell some fifty years later could only achieve a grey or off-white using a much improved version of Young's coloured disc method. That Young should have inferred the formation of white light in such experiments is, of course, reasonable; inference drawn from experiment is the very essence of science especially when such conclusions can be confirmed from other experiments as well. Here too Young was seemingly successful, strengthening his conclusion by still another colour mixing method. He stated

> If we mix together, in proper proportions, any substances exhibiting these colours in their greatest purity, and place the mixture in a light sufficiently strong, we obtain the appearance of perfect whiteness; but in a fainter light the mixture is grey, or that hue which arises from a combination of white and black. [14, vol I, p345]

Once again we have 'perfection', this time by a method of colour powder mixing described by Newton. Indeed, it seems from his description that this demonstration was visually more striking than the disc. But there is an irony here in Young's finding support from these two types of experiments, for the spinning disc method produces a new colour by successive superposition of different colours in the eye resulting in a new colour sensation not originally present. In the powder method, colours already present in the light are removed by filtration thereby enhancing the effect of other colours also present but also lowering the overall brightness. Mixing colours by the spinning disc method is additive colour mixing as the superposition of the colours would imply. Mixing colours by pigments is referred to as subtractive mixing. This distinction, however, was not clarified until 1852 when Hermann von Helmholtz showed that neither method could give supporting evidence to the other. Young, of course, could not have known this and drew what seemed to be reasonable conclusions. He apparently viewed his 'perfect whiteness' under bright sunlight. Although there would have been much absorption, still enough of the sunlight colours would have been reflected from the powder mixture to give a white sensation through a dazzling effect—especially if the pigment mixture was not a deep one. From such crude observations and with a little imagination Young easily could have persuaded himself of a result he had already determined. Then viewing the mixture under subdued light (such as a cloudy day) he naturally could have expected a lower intensity of reflection or a grey sensation coinciding nicely with the disc. In this Young was hardly doing more than Newton and Robert Boyle before him. (See Chapters 4 and 5 for a history of this confusion.) How fortunate he was to have rejected his original choice of primary colours. Had he tried to make white from red, yellow and blue by the disc method he would also have produced a grey. Had he then attempted the powder method of combination with these colours he could also have achieved white through a dazzling effect under strong sunlight. Indeed Newton had done similar experiments. With these crude techniques he could have argued for either case, as was indeed done by his successors. Here was truly a case where two wrongs did make a right.

These events have shown how a series of brilliant speculations and even brilliant errors led to an important idea. Young's original hypothesis in the Bakerian Lecture concerning colour vision had absolutely no anatomical or physical basis. In fact, the whole idea was actually irrelevant to his real objective of introducing the wave theory of light together with his highly original concept of interference, and it played no further role in his paper. Similarly the faulty spectral interpretation from which he amended his choice of primary colours was relegated to an insignificant portion of yet another paper also aimed at establishing the wave theory of light. Even the concept of three primary colours itself was only a popular but unproved notion. Perhaps the genius involved here was that Young was able to collate materials from such disparate sources and push the sound–light analogy far beyond the limits his predecessors had reached. As Young himself once remarked of his role in science, 'acute suggestion . . . was always more in the line of my ambition than experimental illustrations.' [15] If this theory of vision had little observational basis, it did have Young's wide-ranging knowledge which made possible an acute suggestion. That it should have captured the imagination of so many people afterwards and that it should have continued to serve them as a working hypothesis is the testimony to Young's genius.

YOUNG ANTICIPATED

Strangely enough, Young's theory of colour vision was actually anticipated before his paper of 1802. In a work entitled *Theory of Colours and Vision*, George Palmer wrote in 1777 that all colours of light are actually composed of three primary colours only, red, yellow and blue. [16, p40] Applying this idea to vision, Palmer hypothesised that 'the surface of the retina is compounded of particles of three different kinds, analogous to the three different rays of light; and each of these particles is moved by its own ray.' [16, p41] The motion of all these particles in unison then corresponded to a white sensation while no motion of the particles corresponded to darkness.

Palmer extended that idea slightly in 1786 to say that the 'retina must be composed of three kinds of fibres, or membranes, each analogous to one of the three primary rays and susceptible to being stimulated by it.' [17] He even conceived that a deficiency of the fibres would result in 'false vision', a term he understood to mean altered colour perception or colour blindness. Noting the severe colour blindness of the French poet Charles-Pierre Colardeau he thought, incorrectly, that Colardeau saw only black, white and grey. He therefore speculated that Colardeau experienced a deficiency in all his colour sensitive fibres. The gradual recognition of colour blindness as a visual defect is discussed in detail in Chapter 7 where it is seen that Colardeau was apparently blind to red but did see certain other colours. If so, Palmer's speculation can be understood to mean simply a difficulty with only the red receiving fibres, which is comparable to the modern view of this form of colour blindness. Young too arrived at this idea in 1807. Referring to John Dalton whose vision of red was also severely restricted, Young presumed that Dalton's lack of red perception could be understood most simply as the 'absence or paralysis of those fibres of the retina, which are calculated to perceive red.' [14, vol II, pp315–16] [18] Thus both Young and Palmer drew similar conclusions from closely similar hypotheses.

Palmer, however, came to these ideas by a different route from Young. He thought, again incorrectly, that only three colours of light actually existed so it was then a short step to hypothesise three receptors in the eye corresponding to the presumed colours. Young, on the other hand, did not believe in only three colours of light—at least not when he first proposed his thesis. Young, as noted, had actually calculated a table of wavelengths of light from red to violet using Newton's measurements on thin films. Then considering the physiological effects of all these different wavelengths in producing vision, Young employed the common opinon from painting of red, yellow and blue as primary colours. For Young, as opposed to Palmer, each colour receptor was stimulated by several colours of light not simply by the specific supposed actual hue. As said before, in proposing his theory of vision, Young merged two different interests of long standing (physiology of vision and physics in terms of the sound–light wave analogy) into one comprehensive scheme. Nevertheless Palmer had certainly grasped Young's essential ideas together with their implications many years earlier.

If Palmer anticipated Young's general concepts, Chrétian-Ernst Wünsch in 1792 anticipated his final selection of primaries. [19] From experiments with spectral colours Wünsch, like Palmer, concluded that only three primary colours existed in the spectrum: red, green and violet. He noted correctly that mixtures of red and green (light) produce orange and yellow, a fact which went completely unnoticed for another sixty years. Wünsch also observed that mixtures of green and violet tended to yield blue and indigo.

Hence although he did not state explicit reasons for his choice of red, green and violet he apparently selected two colours at the ends of the spectrum and one in the middle. As contrasted with Young, Wünsch arrived at his choice of colours for much better observational reasons. However, Wünsch's opinion as to three spectral colours contradicted Newton's views of seven primary spectral colours. In a French translation of his work, the respected physicist–meteorologist, Hassenfratz, found much to criticise in Wünsch's observations. While admiring his experiments, Hassenfratz could not accept anything violating accepted Newtonian doctrine. Also, Wünsch's selection of primary colours contradicted the conventional wisdom that red, yellow and blue formed the *only* correct choice of primary colours—a point which Hassenfratz again did not miss. So ultimately the opinions of Wünsch and his experimental observations were rejected and forgotten.

Unfortunately, neither Palmer's nor Wünsch's works ever experienced wide circulation and are today exceedingly rare. Subsequent events in this history will show that Palmer, especially, remained unknown to nineteenth and early twentieth century investigators who universally regarded Young as their progenitor.

Indeed Palmer is an enigma; who he actually was and even his nationality remain unknown. The title pages of his works actually show his name as G Palmer and the name George is at best inferred from other evidence. There is strong reason to believe the name G Palmer itself is actually a pseudonym. A delightful and exhaustive attempt to identify Palmer is told by Gordon Walls [20] who traces a vague reference to him in Helmholtz's *Handbuch der Physiologischen Optik* (1867) to the physicist August Seebeck in 1837 (see Chapter 7 on the discovery of colour blindness) and from there to a German popular science magazine edited by L C Lichtenberg and J H Voigt in 1781. While both Seebeck and Helmholtz had undoubtedly read the journal article, neither had seen the original 1777 publication itself, and neither knew of the 1786 work. [21] Suffice it to say that Palmer's works went unremarked by posterity, his identity yet remains a mystery, and his keen insight had no influence whatever on later investigators, Young included. Palmer's two books are not listed in Young's *Catalogue of Works Relating to Natural Philosophy* some 20 000 items long but the Lichtenberg and Voigt article is noted there although slightly miscited. Young probably did not know of Palmer's works. Whether he had actually read the Lichtenberg and Voigt article is a matter of conjecture. Even if he had, considering the evolution of Young's thought traced in this chapter, it seems unlikely that it played any role in Young's ideas. Yet it is a striking fact that the essence of Young's theory of colour vision as well as his final choice of primary colours were both actually available long before his Bakerian Lecture. Through a series of circumstances, appearing too early to be appreciated and lacking sufficient evidence, these early ideas languished, as in fact did Young's theory to a certain extent. Much of the rest of this history narrates, in effect, the changes of context necessary to place these brief thoughts in a new light!

NOTES

1 Priestley, Joseph 1772 *History and Present State of Discoveries Relating to Vision, Light and Colours* (London)

2 Newton, Isaac 1730 *Opticks* 4th edn (London). Reprinted 1952 (New York: Dover)

3 Hartley, David *Dictionary of National Biography* **IX** 68

4 Darwin, Robert Waring 1786 New experiments on the ocular spectra of light and colours *Phil. Trans. R. Soc.* **76**

5 Young, Thomas 1800 Outlines of experiments and inquiries respecting sound and light *Phil. Trans. R. Soc.* **90**

6 Young, Thomas 1801 On the mechanism of the eye *Phil. Trans. R. Soc.* **91.** This paper was the Bakerian Lecture delivered in November 1800

7 Young, Thomas 1802 On the theory of light and colours *Phil. Trans. R. Soc.* **92.** This paper was the Bakerian Lecture delivered in November 1801

8 Newton, Isaac 1672 Mr Isaac Newton's answer to some considerations on his doctrine of light and colours *Phil. Trans. R. Soc.* as quoted in [7, p18]

9 Newton, Isaac 1757 Second paper on colour and light, read at the Royal Society in 1675 but published in Birch, Thomas *History of the Royal Society of London* vol 3 (London) p262 as quoted in [7, pp19–20]

10 Airy, George Biddell 1833 Remarks on Sir David Brewster's paper 'On the absorption of specific rays, in reference to the undulatory theory of light' *Phil. Mag.* Ser. 3 **2**

11 Young's values give extreme red as 676 nm and 424 nm for extreme violet, percentage differences from modern values being about 3% and 6% respectively

12 Wollaston, William Hyde 1802 A method of examining refractive and dispersive powers by prismatic reflection *Phil. Trans. R. Soc.* **42**

13 Young, Thomas 1802 An account of some cases of the production of colours not hitherto described *Phil. Trans. R. Soc.* **42**

14 Young, Thomas 1807 *A Course of Lectures on Natural Philosophy and the Mechanical Arts*. Republished 1845 ed. P Kelland (London). References are to the 1845 version

15 Peacock, George 1855 *Life of Thomas Young* p397

16 Palmer, George(?) 1777 *Theory of Colours and Vision* (London). The remarks here are based on an excerpt from MacAdam, David L (ed) 1970 *Sources of Colour Science* (Cambridge, MA: MIT Press). Palmer's works are very rare; there are only two copies in the United States, one in English and one in French

17 Palmer, George 1786 *Théorie de la Lumière* (Paris). English excerpt in MacAdam, David L (ed) 1970 *Sources of Colour Science* (Cambridge, MA: MIT Press) p48

18 The remark is made in his *Catalogue of Works Relating to Natural Philosophy* as an annotation to the entry of Dalton's paper on colour blindness (see Chapter 7, note [11])

19 Wünsch, Chrétian-Ernst 1792 *Versuche und Beobachtungen uber die Farben des Lichtes* (Leipzig). A long excerpt in French appeared in 1807 in *Ann. Chim.* **64**

20 Walls, Gordon 1956 The G Palmer Story *J. Hist. Med. Allied Sci.* **XI** 66–96

21 To add to the mystery of just who Palmer was, the work of 1777 was written in English and published in London, while the work of 1786 was written in French and published in Paris

2
Spectral Analysis by Coloured Filters and Brewster's Theory of the Spectrum

Young's trichromatic theory of vision, regardless of its origins, placed the question of colour perception on a testable basis. By providing a working hypothesis it raised the question itself to a new awareness. Young thought he had himself confirmed his own theory, at least in a limited fashion, but his experiments were hardly definitive. Similarly Wollaston's interpretation of the spectrum, while not rejected out of hand, certainly left many people uneasy. Still, there definitely seemed to be room for improvement on the basic Newtonian conception of the spectrum. The infrared portion of the spectrum was discovered in 1800 by William Herschel [1] while the ultraviolet portion was discovered in 1801 by J Ritter. Thus suddenly the dimensions of the spectrum were enlarged. So when Wollaston announced in 1802 that the visible spectrum consisted of only four colours instead of seven as Newton had said it seemed just possibly true. Not even the elegant discoveries of Joseph Fraunhofer in 1817 and his announcement of 325 dark lines in the sun's spectrum compared to Wollaston's seven had completely dispelled Wollaston's notions. Although Fraunhofer had achieved a degree of dispersion and resolution far exceeding Wollaston's the question concerning yellow as a compound colour in the spectrum was still outstanding some twenty years after its original proposal and seven years after Fraunhofer's work. Therefore it appeared that there was yet a great deal to learn about the solar spectrum and allied topics one of which concerned the primary colours and their possible role in both the spectrum and vision. To this end, David Brewster and John Herschel in 1822 still found it necessary to comment on—and disprove—the spectral interpretation of Wollaston and Young.

In order to investigate the possible existence of primary spectral colours it was really necessary to have so-called 'pure' colours which could be accomplished by isolating narrow portions of the solar spectrum by slit arrangements. But if yellow light really had no independent existence in the spectrum, as seemed just possible, such an arrangement would be of doubtful validity and probably self-defeating. Clearly what was needed was a new technique for producing monochromatic light from the spectrum together with a deeper understanding of the spectral structure itself. In related problems astronomers had long been using coloured or smoked glasses to attenuate the light through their telescopes. William Herschel described their use in telescopic observation in his famous paper of 1800 on infrared light. Coloured glass spectacles had also been prescribed for people with 'tender vision', [2, p102] those whose eyes were particularly sensitive to bright light. Truly they viewed the world through rose coloured glasses—as well as through greens and blues. [2, p102] Indeed, prevailing choices in colours changed with the fickleness of fashion, which fact first attracted David Brewster's attention to coloured filters. Tongue-in-cheek, Brewster wrote that this fickleness 'unless checked by the application of some principle, may soon carry us through all the colours of the spectrum.' [2, p102] So Brewster set out to find the required principle for prescribing coloured spectacles.

Now Brewster at this time was 41 years old and at the peak of a fruitful career in optical research. Having published many important papers on polarised light, he had discovered the notable new phenomena of elliptical polarisation and double refraction in crystals along two axes. Prior to this, double refraction had been known only along a single axis. This discovery he then employed to classify several hundred minerals and crystals into specific groups by their optical properties thereby creating the new field of optical mineralogy. More popularly he became famous in 1816 for his invention of the kaleidoscope which resulted directly from his researches in polarisation by reflection. Receiving many honours for his work, he reaped a double portion in 1815 when he was both elected a Fellow of the Royal Society and awarded the Society's most distinguished honour, the Copely medal, for his 'paper on the polarisation of light by reflection.' It is this paper which states his famous law relating the index of refraction to the angle of polarisation of light. This award was followed by the prestigious Rumford medal of the Royal Society in 1818 again for his discoveries relating to polarisation and ultimately he received the Society's Royal medal in 1830. In short, Brewster was then one of Europe's leading optical researchers, which he accomplished while actually earning his living by voluminous literary output and scientific editing.

The researches taken up in 1822 represented a new line of investigation for Brewster. Leaving polarised light, he now turned to analysis of light by absorbing media. He first found that looking through a piece of blue glass caused him quite considerable pain. Just what he viewed through the glass is uncertain but careful experimenter that he was, he immediately set out to find the cause of his pain, this time using a candle instead of the original, and presumably much brighter, source. In front of the candle he placed a prism and between the prism and his eye he placed the blue glass. He thus viewed the candle's spectrum through the glass. Expecting, no doubt, to find a homogeneous blue light, he found the glass had instead

> the remarkable property of absorbing only the middle rays of the spectrum, viz, the *Green*, *Yellow* and *Orange*, and transmitting the

Sir David Brewster
(courtesy of Burndy Library).

> *Violet* and the *Red*. The spectrum therefore, consisted of two separate images, the one *Red* and the other *Blue*; and hence the eye was not able to see distinctly by means of rays of such different refrangibilities. When it tried to adapt itself to the blue rays, it became incapable of converging the red ones; and when it endeavoured to converge the red light to focus, it lost the power of converging the blue. The effect, in short, was the same as if it had attempted, by its muscular power, to adjust itself to two different distances at the same time, and therefore it became completely exhausted with its fruitless efforts to obtain distinct vision. [2, p103]

Brewster had made a fundamental discovery: coloured glasses do not actually transmit monochromatic light, a 'fact' previously taken for granted. Equally fundamental and remarkable was his graphic description of focusing the eye for different colours; he had done nothing less than demonstrate chromatic aberration in the eye.

These two discoveries, simple but elegantly done, solved Brewster's problem of rationally prescribing coloured glasses for people with light sensitive eyes. He required a relatively narrow band of wavelengths of relatively high luminous intensity. To this end Brewster investigated differently coloured glasses until he found a yellow-green type fitting these criteria. It absorbed most of the red and blue light leaving essentially the centre portion of the spectrum for vision. Thus attenuation of the light was effected while difficult focusing due to chromatic aberration in the eye was reduced by eliminating the extreme ends of the spectrum. It was a nice piece of work although one wonders about the ability of glass technology then to consistently produce the yellow-green glass he recommended.

The rest of this initial excursion into filters was devoted to their application to producing monochromatic light for microscopes. It was still not possible in 1822 to make satisfactory achromatic microscope objectives so the search for a monochromatic light source was related to the problem of improving microscope images by avoiding chromatic aberration. Although he arrived at no definite conclusions, his next paper on filtered light was devoted largely to this problem. [3] Suffice it to say of that work that Brewster, by trial and error with different coloured glasses and liquids, achieved a thorough empirical knowledge of the absorption and transmission characteristics of coloured filters. Employing his new skills, he was able to isolate almost any portion of the spectrum he wished thereby providing reasonably monochromatic light for microscopy. This technique he then applied to that old question raised by Wollaston: Does yellow light have an independent existence in the spectrum?

Brewster set up an experiment in which he 'obtained a spectrum exactly similar to that described by Dr Wollaston and Dr Young.' [3, p441] The spectrum was apparently that described by Young and not that of Wollaston of 1802. Wollaston's question was rephrased: Do red and green really combine to give yellow in the spectrum? Brewster observed this spectrum through a red absorbing glass and a green absorbing glass together. Expecting that both red and green would be absorbed then, if yellow were really a combination of the two, it also would have to disappear. The remaining spectrum therefore would consist of only blue and violet. But Brewster saw something quite different. Both red and green did indeed disappear but not the yellow. The narrow band of yellow not only remained but appeared to spread over the spaces vacated by the missing red and green. [3, pp441–2] Therefore he saw a spectrum consisting of yellow, blue and violet with yellow occupying over half the length of the spectrum. From this observation Brewster concluded that spectral yellow is not a mixture of red and green.

It is interesting to compare these results with John Dalton's description of his colour blindness (see Chapter 7). Dalton had shown in 1797 that he could see no red and that his yellows seemed to include the band of red, orange, yellow and green of normal vision. In other words, from red to green in the spectrum as seen by normal vision, Dalton saw one colour only, yellow, in varying intensities. It seems that Brewster had actually simulated for himself Dalton's colour blindness. However, although he knew of Dalton's perceptual difficulties he did not think in these terms preferring to regard these observations differently. Nor did he realise that if Young and Wollaston had been correct then elimination of either red or green alone would have eliminated yellow from the spectrum.

In any case Brewster was jubilant, for he was convinced that his observation showed without doubt that yellow does indeed exist independently in the spectrum as he intuitively assumed it should. Young and Wollaston had been wrong. But more than that he said that this single observation proved most convincingly that 'the Red and Green spaces, at the place where they come in contact, consist of Red and Green rays, respectively mixed with yellow rays of the same refrangibility.' [3, p442] What a remarkable conclusion: red, green and yellow light all of the same indices of refraction! It violated accepted optical observation dating from Newton. Indeed why would chromatic aberration exist if red, green and yellow were all equally refrangible? Still if the yellow light occupied the red–green sections as he thought he observed then one can understand Brewster's rationale. Light from each different section of the spectrum exhibits a different index of refraction, so the yellow must have a range of indices of refraction (for a given material) from the red to the green. On this Brewster said

> The *Yellow* light, therefore, of the solar rays, instead of occupying a separate space in the spectrum, has its most refrangible rays mixed with *green* light of equal refrangibility, and its least refrangible rays mixed with *red* light of equal refrangibility. [3, p442]

From a single observation came a whole new idea. Not only did yellow enjoy its independence, it even usurped the space generally occupied by other colours. And it did so with a variable index of refraction. It is difficult to understand how a man of Brewster's acknowledged competence could conceive different colours of light as having the same index of refraction. Yet he was quite definite about the observation above; being convinced of what he saw, the conclusion followed. On first thought it might appear that he used too wide a slit to form his spectrum thereby causing overlap of several different colours. But this was not the case as subsequent events will show nor can it be reconciled with Brewster's experimental skills. Could it be that Brewster was actually an adherent of Young's concept of three primary colours (for vision) and was now extending that idea to the spectrum itself? This is apparently so and we shall see that he followed this line of reasoning to a perfectly reasonable, ostensibly plausible but ultimately impossible conclusion. Before doing so, however, we shall briefly mention similar experiments performed by John Herschel.

Within the trend-seeking methods for obtaining monochromatic light, Herschel performed filtering (or absorption) experiments. Investigating many specimens of glass and liquid he too showed (in 1823) that yellow is an independent spectral colour and, like Brewster earlier, he could isolate almost any portion of the sun's spectrum he wished. [4, p445] However, unlike Brewster, Herschel, the mathematical physicist, also provided a quantitative expression for the nature of filters. He showed that the brightness of any emerging hue would depend on an absorption ratio for that colour and the thickness of the absorber. In essence then to produce a beam with a 'narrow range of refrangibility' [4, p445] it was only necessary to find filters of suitable transmission characteristics and thicknesses. He illustrated this idea with a lovely example comparing the water of Lake Geneva with that of Lake Como: in small thicknesses both appear colourless, but in considerable thickness Geneva water is indigo-blue while Como water is emerald-green. For Herschel, there were no colours of equal indices of refraction and beyond his little expression for filters he offered no theoretical overlay to his results. A rigour was imposed on his results that was absent in Brewster's so in the end he achieved a clearer understanding of filters than his colleague. In any case Herschel went on to other activities while Brewster returned to this work in 1831 with a whole new conception of the spectrum. For a man not usually given to theoretical explanations, Brewster made a major break with his own practices and then looked to Herschel, partly at least, for support.

Redescribing his first observations of 1822, Brewster in 1831 now said that he used a purplish-blue glass although originally he had stated that it was simply blue. Either he was now giving more information than originally or his memory was fallible; we shall never know. But his new description of the observations also varied from the original. In some detail he gives an analysis of his results without ever describing the actual observations on which the analysis is based. We are left in the dark so to speak. Brewster stated his 'results' as follows:

> This glass absorbed the blue rays, which when mixed with the yellow, made green, and the yellow rays, which when mixed with the red, made

orange; and by insulating the yellow and red, it thus effected a perfect analysis of the compound green and the compound orange. [5]

More simply, the passage says only: ‘the glass absorbed the blue rays . . . and the yellow rays.’ However, Brewster was also stating inferentially that blue light mixed with yellow gave green, and yellow light mixed with red gave orange. Since no blue or yellow were seen there was no green or orange either. Hence the only colours left to view were red and violet, a conclusion corresponding to his 1822 statement about the difficulty of focusing for red and violet together.

Having produced an analysis of the ‘compound green’ and the ‘compound orange’ Brewster once again drew the conclusion that ‘yellow light has an independent existence in the spectrum; and that the prism is incapable of decomposing that part of the spectrum (of four colours) which it occupies.’ [5, p124–5] Let us compare this conclusion with Wollaston’s interpretation of the spectrum. Now Brewster had originally aimed in part to show that Wollaston and Young were wrong about yellow being a combination of red and green in the spectrum and to his own satisfaction he had succeeded. However, his parenthetical remark does not mean that he thought the spectrum consisted of four colours. It only means that yellow when present spread out through the region occupied by the four colours, red, orange, yellow and green. Since $Y + R = O$ and $Y + B = G$, according to Brewster’s notions, spectral orange and green could occur only if the yellow spread over the red–green region. And since he had absorbed the yellow these colours, except some extreme red, were perforce absent. This follows from the known ‘fact’ that $Y + R = O$. If one conceives the red band to extend beyond the space we usually ascribe to it, then yellow rays also extending towards the red would mix with the red giving orange. Significantly, orange was absent when viewed through the purplish-blue glass since the yellow had been absorbed. Similarly one can conclude that certain blue and yellow rays have equal refrangibilities. Not surprisingly, Brewster criticised the accepted notion of seven spectral colours pompously stating that ‘this unequivocal result of a simple experiment at once saps the foundation of the prismatic analysis of light.’ [5, p125]

Indeed these ideas lead to an interpretation of the spectrum in terms of three separate, equally long, overlapping spectra of primary colours. Since $O = R + Y$ and $G = B + Y$ according to received opinion at that time the implication is that red, yellow and blue are primary spectral colours forming three separate and distinct overlapping spectra. Following Brewster the spectrum can be illustrated as in Plate 3. [5] The abscissae are the spectral colours; the ordinates represent relative light intensity. Therefore red has a distribution of indices of refraction and is visible throughout the entire length of the spectrum. Similarly, both yellow and blue, each with variable index of refraction, are visible through the entire length of the spectrum. Observed spectral colours are then actually combinations of the separate primary spectra shown in Plate 3. Each colour has a maximum intensity in that part of the spectrum where it is usually seen but each also extends beyond those regions in sharply diminished intensity. Ultimately, each primary spectrum can be separated from the others only by the new method of filtering.

Brewster gave an example of how the primary colours produced other spectral colours. A colour at position X in Plate 3, figure 4 can be given a simple algebraic interpretation. Remembering that the ordinate represents the light intensity, we arbitrarily let

$$\begin{aligned} \text{red} &= ax = 30 \text{ rays} \\ \text{yellow} &= bx = 16 \text{ rays} \\ \text{blue} &= cx = 2 \text{ rays} \\ \hline \text{total} &= ax + bx + cx = 48 \text{ rays of light} \end{aligned}$$

and the 'point X will be illuminated with 48 rays of light, i.e. 30 of red, 16 of yellow, and 2 of blue.' [5, p127]

This now leads to another way of looking at the example. Whatever the primary colours may be they must recombine into white light in proper proportions. This much, at least, of Newton's analysis was preserved. Brewster assumed, again arbitrarily, the proportions of red, yellow and blue (in intensity) required to make white light to be 3, 5 and 2 respectively. Then subtracting these amounts from those above, but adding their total value back again as white, to maintain the original intensity, the colour at X may be regarded as being composed of

$$\begin{aligned} \text{red} &= 27 \text{ rays} \\ \text{yellow} &= 11 \text{ rays} \\ \text{white} &= 10 \text{ rays} \\ \hline \text{total} &= 48 \text{ rays.} \end{aligned}$$

We conclude that the two blue rays contribute no bluish tinge to this point, as they must not for we know point X is orange. Hence Brewster concluded that the colour at X is orange 'rendered brighter by a mixture of white light.' [5, p128] Given the colour X in Brewster's example we know it is composed of three separate primary colours all of equal index of refraction. Therefore they *cannot* be separated by a prism; but they *can* be separated by passing the light through absorbing media or filters.

Suppose then we transmit the light through a medium which absorbs the 27 rays of red. (In principle, this is possible.) The light emerging must consist of 11 rays of yellow and 10 rays of white; it is a brilliant yellow, according to Brewster's remark above that the orange was rendered brighter by the white. Now transmit the light through another medium which absorbs the 11 rays of yellow. The light transmitted or emerging must be white! Not only that but it must be a new species of white light, a 'monochromatic' white undecomposable by a prism and heretofore undiscovered. Small wonder that Brewster could proudly write:

the existence of such light has never even been conjectured, and its

> insulation at any point of the spectrum becomes a proof of the existence, at that point, of red, yellow and blue rays of equal refrangibility. [5, p129]

What an amazing conclusion. Now one can understand Brewster's previous remark that these results sap the very 'foundation of the prismatic analysis of light.' For the very basis of that analysis is Newton's famous remark that 'to the same degree of refrangibility ever belongs the same colours, and to the same colours ever belong the same degree of refrangibility.' [6] But Brewster now claimed that this was not necessarily so. Here each colour extends throughout the spectrum varying its index of refraction while yet maintaining the same hue. The separate variations in index of refraction of the primaries will always just compensate so all three are exactly equal at any particular point in the spectrum. Therefore although we can measure different indices of refraction throughout the spectrum this new white light will be an exception. This 'white light' cannot be white in the Newtonian sense for it is unresolvable by a prism. It can be analysed into only three colours and then only by filters; let us call it 'Brewster's white.'

Brewster's 'discovery' and theory seriously challenge the concept of chromatic aberration. Although it does not forbid such a phenomenon, clearly the Newtonian conception would need major modification. Possibly, even, the new white light was a different kind of light from that described by the new undulatory theory and as such it would pose serious difficulties for that theory. There a major unsolved theoretical problem concerned the very functional relationship between index of refraction and wavelength. Brewster's discovery would add to this difficulty that of understanding index of refraction in terms of brightness as well if, of course, the wave theory were applicable at all to the new light. About the only Newtonian idea preserved by Brewster was that his primaries must still give white when mixed in proper proportions. Still, if Brewster's discovery were correct it would be a spectacular one indeed, rivalling Newton's own analysis of the spectrum and ensuring immortal fame for its author. Small wonder that Brewster considered this his finest work and, like Newton with his early optical papers, he stoutly defended the work against all detractors.

For all this one wonders whether Brewster ever actually tried the procedure above. Did he really think he had observed 'monochromatic' white light? He never explicitly stated that he did observe it and his experimental description and observations are often vague. Certainly the quotation above strongly implies that he really isolated such a light—to his own satisfaction—while his own high integrity and devotion to experiment would prevent him from speaking from conjecture only. Although he was now writing mainly from memory he undoubtedly believed he had seen his 'Brewster's white.' What he actually viewed, very likely after passing the light through several filters, was a very dim light giving the prism face a feeble overall illumination. Then if the brightness were so low as to shift his vision from colour to black and white perception he could presume he was viewing an unresolvable white. It would be analogous to Young interpreting the grey on his spinning disc as really white and, like Young, he could easily extrapolate from this low-level grey to white. Further thoughts on what he actually might have seen are discussed in Chapter 3.

Before discussing Brewster's conceptions further let us summarise them here:

(a) The primary colours of the spectrum are red, yellow and blue. These combine in appropriate proportions to make the other spectral colours and in proper proportions will also give white.
(b) The three primary colours form separate, overlapping spectra of equal length.
(c) The index of refraction of each primary colour changes throughout the spectrum but at any one point in the spectrum the three colours have equal indices of refraction; therefore, they cannot be separated by a prism. As a result a certain quantity of 'monochromatic' white light exists at every point in the spectrum adding to the brightness of the colour. This white may be isolated from the other colours only by absorption methods.

In announcing his radical ideas Brewster also overcame an objection to them that had arisen even before their formal publication. We have seen that he had earlier hinted at them in 1823. Returning now to those observations he invoked some remarks of John Herschel for support. He stated that Herschel had originally embraced his ideas in reporting his own absorption observations. This is certainly a generous interpretation of Herschel who then showed neither acceptance nor rejection of Brewster's concepts. Herschel did, however, write in 1827 that there was a certain plausibility to these arguments. [7, p432] The spectrum viewed through absorbent media he stated lost all appearance of a 'regular gradation of colour from one end of the spectrum to the other.' [7, p434] The space between red and yellow could be made very small by choosing a suitable thickness of glass. When such a small space was achieved, the transition between red and yellow became quite sudden, the orange completely disappearing. Herschel asked, 'what then . . . is become of the orange; and how is it, that its place is partly supplied with red on one side, and yellow on the other?' [7, p434] Without answering the question, Herschel wrote

> these phenomena certainly lead us very strongly to believe that the analysis of white light by the prism is not the only analysis of which it admits, and that the connection between the refrangibility and colour of a ray is not so absolute as Newton supposed. [7, p434]

This is, however, the strongest support, if support it is, Herschel ever gave Brewster. Indeed he even qualified himself, noting that there were certain objections to Brewster's theory based on data from persons with defective colour vision. Colour blind people presented with optical hues of known composition as opposed to the compound colours of painters, distinguished two colours only, yellow and blue. (See further details on colour blindness in Chapter 7.) This defect Herschel had found arose only from a defect in the 'sensorium', not from any optical properties of the eye nor possible colouring matter in the humours of the eye. [7, p434] [8]

Brewster was annoyed at Herschel's alleged desertion from his views. 'An obscure physiological fact occurring in one eye out of a million, could not, on any principle, affect the result of a legitimate induction' according to Brewster. [5, p126] He could answer Herschel's objection by stating simply that the eyes of colour blind people are insensible to red light. [5, p135] For according to his scheme if we remove the red light from a spectrum, there are left, of course, only yellow and blue; and these are exactly the two colours which such people see, according to Herschel. Moreover, since the red, yellow and blue spectra overlap, persons insensitive to red will still see red objects but of a different colour—green, since $G = Y + B$. Additionally, blue light is seen in the violet section of the spectrum and a greenish-yellow appears in the orange and red sections. For such people the spectrum consists of only two colours, yellow and blue. Thus said Brewster, 'the physiological fact, and the optical principle, are therefore in perfect accordance; and while the latter gives a precise explanation of the former, the former yields to the latter a new and unexpected support.' [5, p135] With this remark we have come full circle. Brewster, like Young, had discovered a correct fact about colour vision but for the wrong reasons and certainly from questionable observations. He had beautifully overcome Herschel's objection by bringing new and different phenomena to support the same conclusions. Since it is ever the aim of scientists to make their theories as general as possible such arguments are very powerful. Indeed a few years later Brewster found another application of his theory thereby generalising it further. The application was partly a negative one yet still one to be reckoned with. In the ongoing discussions at the time over the undulatory versus emission (particle) theories of light the champions of each side seized upon any new optical discovery as a token of favour to their cause. Brewster, outspoken champion of the emission theory, who would not 'kneel at the new shrine' of the wave theory [9, p361] employed his triple spectrum to argue against the undulatory theory although admittedly it did not argue in favour of the emission theory.

In 1833 Brewster argued that the phenomena of absorption of light 'militated strongly against the undulatory theory.' [9, p363] He objected mainly to the requirements imposed upon the ether medium by selective light absorption. He pointed out, for example, that an absorptive substance he had used was absolutely opaque to violet light but passed red freely. Hence on the basis of the wave theory one must suppose that the ether contained in the substance was set into undulation by the red rays but was immovable when acted upon by violet light which differed from the red only in terms of wavelength. Conversely there were substances opaque to red and transparent to violet. Even worse perhaps, oxalate of chromium and potash passed red light whose index of refraction in flint glass was 1.6272 and another red whose index in flint glass was 1.6274 but it would not pass red of index (in flint glass) of 1.6273. Of this state of affairs Brewster wrote:

> Among the various phenomena of sound no such analogous fact exists, and we can scarcely conceive an elastic medium so insularly constituted as to exhibit such extraordinary effects. We might readily understand how a medium could transmit sounds of a high pitch, and refuse to transmit sounds of a low pitch; but it is incomprehensible how any medium could transmit two sounds of nearly adjacent pitches, and yet obstruct a sound of an intermediate pitch. [9, p363]

Since the origins of the undulatory theory were so interwoven with analogies with sound Brewster doubtless felt justified in continuing that tradition here. But in doing so he was out of touch with the times. The earlier analogies had argued only the plausibility of an undulatory theory of light without giving it a formal mathematical framework. By 1833, however, the wave theory had considerable mathematical foundation and this argument was really passé. Indeed this objection and the one following were never taken too seriously by his colleagues.

Experiments with nitrous acid had shown Brewster that it selectively absorbed light when in the gaseous state but passed all light when in liquid form. Since the ether of nitrous acid would presumably exist in a freer state when the substance was gaseous, this situation was just the opposite of that expected. Hence, Brewster could hardly allow philosophically the existence of a luminiferous ether and so felt impelled to support 'the falling temple in which Newton once worshipped.' [9, p361]

It was a strange state of affairs. Brewster who was attempting to overthrow Newton's most fundamental discovery regarding light was also attempting to preserve his corpuscular theory of light. With the wave theory enjoying increasing acceptance, Brewster was fighting a rearguard action. His attachment to his triple spectrum, besides that to be expected, was also related to his stubborn rejection of the wave theory of light. In fact, in a personal remark to John Tyndall, Brewster said that he could not imagine the Creator capable of so clumsy a contrivance as filling all space with an ether. [10] He found the undulatory theory ultimately inconceivable and the emission theory, which he ascribed to Newton, to be the correct one.

The elastic properties of the ether actually did present major problems to the wave theory in this period but other physicists took a more constructive view by seeking solutions to those problems rather than destroying an otherwise successful theory. Brewster on the other hand could not separate his physics from a certain religious sense and was mildly chastised for this (posthumously) by Tyndall who replied that Brewster was on very dangerous grounds professing to know too much about the mind of the Creator. [10]

Brewster's arguments above were deftly answered by the Astronomer Royal, George Biddell Airy, [11] but Brewster maintained his opinion and in 1837 again employed his triple spectrum to argue against the wave theory. He compared the experimental similarities and differences between absorption phenomena and the colours of thin plates. [12, p245] As before, Brewster pointed out that the phenomena of absorption were inexplicable by the theory of colours of thin plates, or in other words, the wave theory. They were however inexplicable by the emission theory as well and it was to show that a theory of light absorption was wanting altogether that he wrote this comparison. [12, p245] Here his concept of spectral colours of equal refrangibilities played an important role as did also the 'non-prismatic' white light he believed he had observed. Speaking of the coloured bands seen through a smalt-blue glass, Brewster wrote

> the *colours of some of the luminous bands are not the same as those of the spectrum*, and therefore the glass has removed certain colours while it has left others of the same refrangibility. The *green*, for example, is changed into *yellow* by the removal of blue rays, and in certain glasses a band, *almost white*, is produced. The colours thus removed are said to be absorbed; and by an extensive series of experiments with such absorbing substances I have been able to insulate white light in the spectrum, which no prism can decompose, and to establish the existence of three equal and superposed spectra of red, yellow and blue light. [12, p248]

Contrasting thin-plate colours with absorption colours, Brewster found that there was a remarkable point of difference between the two classes of phenomena which was yet unexplained. By implication he required the wave theory to explain more than just absorption. It would also have to explain the *change* of colour by absorption (as he believed his experiments showed) and the equal refrangibilities of light of different colours. Then the appearance of his 'non-prismatic' white light would also have a theoretical basis. Since nothing indicated the possibility of success in this direction, Brewster once again thought this was an argument, albeit a negative one, against the wave theory. Admittedly, however, his experiments likewise had no explanation within a corpuscular framework.

Ultimately Brewster's theory and observations were so deeply interwoven as to be inseparable. The theory rested basically upon one class of phenomena (absorption) and several tacit assumptions. Viewing the spectrum through filters he saw apparent changes in the transmitted colours as compared to their unfiltered appearances. He was particularly impressed with the case of absorption of orange and green and the juxtaposition of red and yellow in the filtered spectrum. Moreover he assumed, with everyone else of his time, that there were three objective primary colours and without saying so he also subscribed to Young's adaptation of this assumption for purposes of vision. Disagreeing with Young on the choice of primaries, in effect he adopted Young's basic concept of three-colour vision and went on to elaborate that theme. But we may ask: Were not other absorbing media available, absorbing other colours? Were there no other colour juxtapositions in the spectrum as a result? Certainly the same method could have been employed, in principle, to argue for a spectrum of primary colours other than those chosen by Brewster.

A similar point was raised by Herschel in 1827 when he stated with regard to prismatic colours 'any three prismatic rays may still be equally assumed as fundamental colours, and all the rest compounded from them.' [7, p437] Nevertheless received opinion held red, yellow and blue as absolute choices which Brewster accepted completely. As for other absorbers he was indefatigable, but in the end he was most impressed with results from the blue glass, undoubtedly because it correlated so well with his preconception of primary colours. Then observations through other media supported these ideas. If Herschel was lukewarm to the theory others regarded it more positively. The Royal Society of Edinburgh awarded Brewster its Keith Biennial Prize in 1834 for this work indicating that it was highly regarded by his peers, all of whom of course were subject to the same tacit assumptions. In the end the theory so relied on an interlocking series of assumptions and ostensibly reasonable observations that contradiction of any one part of it tended to upset the entire concept. One had to accept it completely or not at all. Fourteen years were to pass without criticism before four leading scientists of the time, Sir George Biddell Airy, John Draper, Marcedinio Melloni and Hermann von Helmholtz, saw the need to dispute Brewster's conclusions. In raising a dispute over the primary spectral colours Brewster stimulated work in others which we shall see eventually led to a clear understanding in 1852 of the difference between colour combinations of light and pigments. We shall now look into Brewster's theory as viewed by its critics.

SIR GEORGE BIDDELL AIRY AND WILLIAM WHEWELL: COMMENTS ON BREWSTER'S SPECTRAL ANALYSIS

Among those who thought that Brewster's conclusions were doubtful were two luminaries of nineteenth century science, Sir George Biddell Airy, the Astronomer Royal, and William Whewell, philosopher, historian of science and Master of Trinity College, Cambridge. The attack on Brewster's theory was opened by Whewell in 1847 in his famous work *History of the Inductive Sciences*. Here he mentioned that Brewster's theory of the spectrum constituted a modern objection to Newton's theory of refraction. [13, p394] Whewell considered Newton's doctrine of dispersion as completely established by experiment insofar as refraction was concerned. But as to any other kinds of analysis, such as that by absorbing media, Newton's experiments did not apply; therefore one could not say that the spectrum was not capable of Brewster's type of decomposition. For Whewell, the cautious commentator, the situation regarding the validity of Newton's law of refraction versus Brewster's theory was unsettled. With much justification he wrote that 'the whole subject of the colours of objects, both opake and transparent, is still in obscurity.' [13, p394]

The latter remarks are contained in Whewell's main narrative and are reasonable enough. It was in a footnote that Whewell initiated the attack on Brewster. Here he stated frankly that the entire analysis was rather arbitrary including both the choice of three elementary colours as opposed to some other number, and the particular colours chosen. Surely, thought Whewell, Brewster could have obtained the same analysis with any other elementary colours. Moreover, did Newton's experiments prove nothing or were they applicable only to light not analysed by absorption? Finally, and most tellingly, he appealed to the authority of the great G B Airy whom he thought had repeated Brewster's experiments without success. [13, p504] Therefore Whewell wrote forthrightly that:

> After a careful consideration of Sir D Brewster's asserted analysis of the solar light into three colours by means of his absorbing media, I cannot consider that he has established his point as an exception to Newton's doctrine. [13, p504]

The dispute was now on, for this comment (just a note to a long narrative) stimulated Airy to elaborate his opinion of Brewster's theory since his name had been mentioned by Whewell.

Airy now thought it incumbent upon himself to give the public a detailed description of the experiments alluded to by Whewell. [14, p73] At the outset Airy, also in 1847, stated that given the 'alleged experimental facts' of Brewster there was still insufficient reason for choosing red, yellow and blue as primary spectral colours; other colours of the spectrum would probably serve just as well. [14, p74] In fact, the entire problem revolved about only one crucial question: 'Is it established that the colour of any portion of the spectrum is changed by the use of any absorbing media?' [14, p74] To answer this, Airy described a series of experiments he had performed in 1833, but which remained unpublished. He stated first of all that the eye had 'no memory for colours' [14, p74] and that no real decision could be made upon the question above unless two spectra were exhibited side by side; one being the altered specimen and the other unaltered. Then an unhesitating comparison could be made. This method, of course, had not been used by Brewster.

Having lost his records of the experiment, and frankly quoting from memory some 14 years afterwards despite his disclaimer of the eye having no memory for colours, Airy described his procedures. In a darkened room where no extraneous light was admitted he cut a hole in one shutter admitting sunlight. The pencil of light admitted was then incident on a metal plate with a narrow slit and the beam of light passed through the slit and fell onto a prism. The resultant spectrum was incident on a screen. Between the hole in the shutter and the slit he placed a holder for the various absorbing media to be tested (see Figure 2.1). By this method he obtained a good spectral image showing several of the stronger Fraunhofer lines. [14, p74] To compare the modified spectrum with an unmodified spectrum the absorbing material was placed in the holder to intercept a portion of the light passing through the aperture so that the two spectra viewed on the screen would then be exactly next to each other and the Fraunhofer lines visible would serve as reference lines. This is indicated by the vertical chain line on the slit in Figure 2.1.

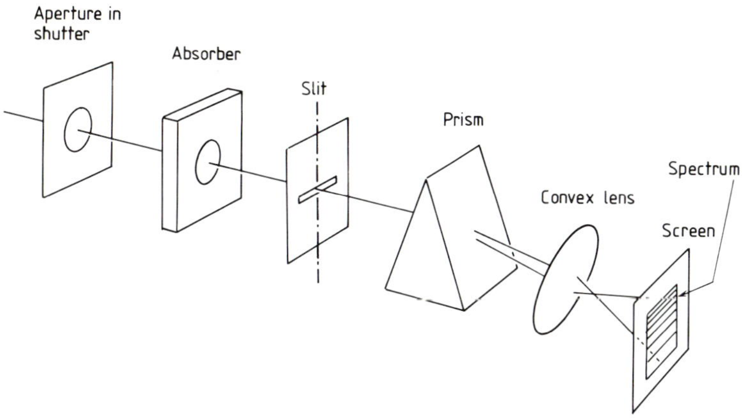

Figure 2.1 Experimental arrangements of G B Airy based on his description aimed at contradicting Brewster's absorption observations.

The left half, say, would be the normal spectrum while the right half would be altered by the filter.

Airy's absorbing materials, like those of Brewster and Herschel, were just about anything available. Unfortunately, Airy was not able to say exactly what the materials were. Contrasted with his precise method above, this lapse gives a poor impression, a point Brewster did not miss in reply. Airy's conclusion, as expected from his purpose of detracting from Brewster's work, was simply that the filters produced no change in the qualities of the colours. This he considered was undoubted, particularly because he viewed the colours in a very dark room, as (he thought) Brewster had not. He objected strongly to Brewster possibly allowing stray light into his room which would certainly affect the colours he saw. Finally Airy decided that no colour extended in either direction from its usual place in the spectrum.

These results, we observe, are essentially null results and as such are somewhat unsatisfying. To look for an expected observation and not see it cannot but raise questions about the skill of the experimenter. On the other hand, if the experimenter's skill is unquestioned, and this would apply to Airy, then one is left in doubt as to the apparatus itself. Is it sensitive enough, for instance, to exhibit the looked for phenomenon? If one answers this positively there is perplexity as to the non-appearance of the phenomenon. In short, a definite observation, a definite numerical result is usually preferred. If the expected result appears one is happy with the theory on which it is based. If a totally different result appears we can at least say our theory needs revision or replacement. But a null result, unless specifically required by theory, does not argue strongly for either case. Therefore, Airy's null results, quite correct in retrospect, did not then mitigate strongly against Brewster's theory although this was his stated purpose.

The history of science contains numerous examples of criticism and defence of various works. But if ever a devastating reply to attack was provided, Brewster's to Airy is a leading example. Brewster's opening salvo set the tone:

> The remarks of Mr Airy in the last Number of this Journal, on my Analysis of the Solar Spectrum, have, I doubt not, given as much surprise to its scientific readers as they have given to me, and this independently of the correctness or incorrectness of the results to which they refer. When a philosopher examines, and pronounces an opinion upon the researches of others, especially upon those which competent judges have recognised as sound, he is bound to repeat the identical experiments which he challenges, with similar apparatus and similar materials; to state the differences which he observes, to inquire into the causes by which such discrepancies have arisen, to establish his own views by new and effective experiments, and to publish his researches in vindication of his charges against his fellow-labourer in science. Mr Airy, however, has not done this; but, as we shall presently see, has followed a course which is as unusual in the history of science as it may be injurious to its progress. [15, p153]

Since Airy's work was never published and could not be called definitive, Brewster, naturally enough, stuck to his original thesis; absorbing media do change the colours of different parts of the spectrum. If they didn't either he (Brewster) had been blind for the last fifty years or the Astronomer Royal was colour blind. Brewster was also insulted by Airy's insistence that direct comparison of spectra was required by the absence of colour memory in the eye. While that may be true he knew:

> that I have a memory for colours, and that the colours are so much committed to memory by fifty year's schooling, that without any immediate comparison I can tell whether or not the green space has become yellower, and whether or not, by certain combinations of absorbing media, I see before me a portion of white light standing close to a portion of red light in the spectrum, like an almond and a cherry stuck together. [15, p154]

And next a little skeleton emerged from Airy's closet. The method of juxtaposition of two spectra that he prescribed was Brewster's own! Airy had seen it demonstrated some 15 years earlier in Brewster's own laboratory where Brewster had used it to compare Fraunhofer lines with absorption lines of nitrous acid vapour. We can imagine Brewster's umbrage at having his own methods turned against him.

Broadside upon broadside, Brewster demolished Airy. His lack of skill, his exclusion of extraneous light, his hodge-podge of absorbing media were all taken up and quashed. Nor did William Whewell escape Brewster's inflamed pen for it was, after all, Whewell who had prompted Airy to write the article. Had Whewell read Herschel's experiments? Probably not. Was not the Royal Society of Edinburgh award recognition of his great work? Apparently not. Had Whewell even read Brewster's original papers on the subject? Manifestly not. As for Airy quoting from memory after losing his notes, Brewster was contemptuous:

> This is the first time in the annals of science that the recollections of experiments have been given to the world. A philosopher in his decline, when his failing sight and his trembling hand are no longer fit for the delicate operations of experimental inquiry, may be excused for calling up the recollections of his manhood in support or in refutation of some exciting speculation; but no apology can be made for those who, with the means and the leisure for repeating their experiments, bring forward their recollections to discredit or to overturn the researches of others who have laboured patiently and successfully in the same field of scientific research. [15, p157]

The net result was that Airy's results and their subsequent interpretation by Whewell were not to be tolerated. Indeed there was one particularly valid objection. Since Airy had viewed the divided spectrum on a screen, he was open to an influence on his retina from both spectra. In Brewster's own experiments

> not a ray of light entered the room but what passed through the narrow slit; the retina was (the) screen, and the absorbing medium was held close to the eye and immediately behind the prism, the only method of obtaining the purest spectrum from a given prism. [15, p158]

While overcoming the objection about extraneous light, that about influence from other colours was important. It certainly was not clear at this time whether there was such an effect or not; nor was it clear whether the retina was a valid screen for viewing colours. There was still room for doubt about the validity of the method, but Brewster had no doubts as he concluded his reply:

> Mr Airy, by inferior methods as I think, but by superior methods as he thinks, recollects having, fourteen years ago, found that these colours were not changed! The Master of Trinity records the last of these results as undoubted scientific truth, and strives to transmit it to the latest posterity. [15, p158]

With this masterpiece of retaliation Brewster swept Airy (and Whewell) from the field. His theory remained unchallenged.

Brewster's deepening estimation of his theory also begins to show in his defence. In his first paper (1831) he tentatively questioned Newton's statement about each colour of light having its own index of refraction but said simply that in the same medium (i.e. a prism) refrangibility is not a complete test of colour. The 'homogeneous white light' he thought he had isolated was a type of light in addition to that which is dispersed by refraction. He was not then setting out to overthrow Newton's work. Now, however, Brewster said outright that Newton's analysis by the prism was perfect only 'so far as it goes', but Newton was incorrect in stating that to the same refrangibility always belongs the same colour. [15, p157] Even more strongly in reply to another criticism, by John William Draper (the well known American chemist), [16] Brewster again stated that his spectral analysis by absorption not only directly controverted Newton's doctrine but it 'absolutely overturns it'. [17] Brewster now felt he had made a most fundamental discovery. Certainly receiving the medal of the Royal Society of Edinburgh encouraged his opinion of its importance. Thus he almost certainly viewed himself here as a kind of crusader against the Philistines. There was more to come.

Shortly after this, the eminent Italian physicist, Marcedinio Melloni questioned not only Brewster's conclusions about the spectrum but also his very technique. [18, p262] Not doubting that Brewster had seen overlapping spectra of red, yellow and blue, he doubted that such a phenomenon really existed. Attributing Brewster's purported observations to faulty experiments, he attempted to duplicate these results by removing the presumed errors. Suppose, he thought, the anterior face of the prism were covered by a screen so as to expose only a narrow 'fissure' parallel to the prism's axis. Would red, yellow and blue then be found in every part of the spectrum? Clearly, he implied that Brewster's observations resulted from exposing too wide a prism face to the incident beam. The purpose of the slit in a spectroscope is to reduce the overlapping of spectral orders. With too wide a slit some colours from one part of the slit will overlap some of those from another part of the slit. So if Melloni's supposition were correct this could have explained Brewster's viewing of overlapping spectra. By reducing the slit width he proposed to eliminate this error. Overlapping could also be the result of viewing the beam too close to the prism face itself. At a greater distance the colours would be separated better. This possible error too he proposed to eliminate.

Melloni repeated Brewster's experiments using what he thought were Brewster's methods but incorporating his corrections. He formed a spectrum from a circular aperture 10 mm in diameter in a metal plate placed in the shutter of an otherwise darkened room. Placing the prism about 15 ft from the window in a position of minimum deviation, he interposed slips of coloured glass between the prism and his eyes. Like Brewster he viewed the spectrum directly rather than projecting it on a screen. The slit modification, for so he thought it was, was effected in an ingenious way. Using a method he had developed in 1843 (and which is still employed in elementary experiments) he covered the anterior face of the prism with India ink. Then

> from the middle of this there was removed a longitudinal space, which ran from end to end of the darkened band, leaving free and uncovered a

> horizontal line little more than a millimetre in width. The slip of blue glass covered only two-thirds of the prism counting from the painted extremity. [18, p272]

With this arrangement Melloni found that yellow did not spread to the red and green sections as Brewster supposed. Instead, the space corresponding to the orange, the narrow space which had so much impressed Brewster (and Herschel too), was 'occupied by a dark zone.' [18, p272] The dark zone was limited by red and yellow produced by the middle part of the prism covered by blue glass. [18, p272] This showed that these colours are independent of this spectrum 'and belong to spectra of elementary layers superior and inferior to the intermediate line.' [18, p272]

From this observation and others, Melloni concluded that Brewster's methods could not be adapted to the conclusions Brewster had reached. He stated

> The method proposed by Sir David Brewster . . . does not therefore appear to me adapted to the end in view; and until it shall have been proved that the colours of a spectrum, *perfectly pure*, change by the interposition of a medium placed near the prism, the change persisting at every other distance, *the existence of different colours in the same transverse element* of the spectrum ought to be regarded as entirely hypothetical. [18, p274]

In short Melloni rejected the entire concept and opted for the original Newtonian doctrine whereby each colour of light possesses its own index of refraction. Contrary to the rather undignified approach of Airy, however, Melloni also paid Brewster a handsome compliment. While he could not accept this particular work of Brewster's he cautioned his audience to remember that Brewster was the author of many beautiful works in optics, all well confirmed by experiment. And so this one mistake, like that of Newton in diffraction, should never lessen the glory that must belong to his optical researches.

Brewster returned to the dispute in characteristic style, firmly convinced he was again unjustly attacked. Replying in kind to Melloni he began by saying:

> The high and well-merited reputation of M Melloni, and the singular confidence which he seems to place in his own results, render it necessary that I should do more than merely pronounce them to be incorrect, and to have no real bearing, even if correct, upon the views which they are brought forward to overturn. [19, pp489–90]

The major objection was that of the slit arrangement for limiting the incident light. Brewster, as noted earlier, did indeed employ a narrow slit in his experiments but it had been cut in the shutter of his window thus combining both the entry of light to the prism and its limitation in one operation. Brewster disclaimed Melloni's notion of his experimental arrangements saying 'I never made such an experiment, and never would have thought of employing a spectrum thus produced.' [19, p490] Indeed the case was just the opposite from what Melloni imagined. Whereas Melloni had cut a circular aperture 10 mm wide in his shutter, he, Brewster, had used one much narrower. [19, p490] While not giving the size of his own slit, he allied himself with both Fraunhofer and Wollaston, noting that:

> The spectrum described by Fraunhofer was obtained from an aperture one-fiftieth of an inch wide; Dr Wollaston used an aperture of one-twentieth of an inch, while M Melloni uses one of twenty-fiftieths, twenty times wider than Fraunhofer's aperture, and eight times wider than Dr Wollaston's. [19, p490]

Given the 10 mm wide aperture of Melloni the separation of the colours would naturally be much less complete than in the studies of Fraunhofer, Wollaston or himself using narrower slits. Then came the 'coup de grace':

> I should have been ashamed of my inexperience had I used such an aperture and such a spectrum as that which Melloni employed; and I can assure him that he has formed a very low estimate of the time and labour which I have devoted to these researches, and of the manner in which they were conducted. [19, p490]

Therefore it was Melloni who was wrong in assuming that Brewster used a wide slit and obtained a complex spectrum. Melloni had, of course, used a slit only 1 mm wide painted directly on the prism itself, the 10 mm aperture being used only to admit light to the room. This fact Brewster conveniently overlooked or ignored. Nevertheless, it does seem that Brewster used quite a narrow slit in the window shutter itself so the original objection of Melloni was unjustified. Right or wrong, Brewster thought that he had disposed of that argument which was Melloni's main criticism. It might have been avoided had Brewster somewhere, sometime, given the dimensions of his apparatus.

Comparing Melloni's other objections to his own experiments, Brewster characteristically squelched each one and concluded, not surprisingly, that Melloni had not read his original paper, nor Airy's comments, nor Brewster's reply. But even if he had, the experimental techniques would still have been unclear. Much of the dispute was really due to want of a good description of his arrangements. One can ultimately piece them together but only as a result of his replies to his detractors, sources which they of course did not have.

In overcoming the objections of Melloni, Brewster also gave a new explanation of Wollaston's original dispersion experiment of 1802. In doing so he again generalised his theory somewhat, something he had accomplished before in overcoming the objections of John Herschel based on colour blindness. Certainly Brewster looked for such applications as a means of strengthening his theory.

Wollaston, we recall, had observed a spectrum of only four colours, red, yellowish-green, blue and violet. Young, repeating these experiments 'with perfect success,' had modified the term yellowish-green to green. What we may ask has happened to the orange and yellow in the spectrum? The composition of the spectrum so described is due, according to Brewster, to a daylight spectrum. It is a spectrum from a blue sky but not of direct sunlight. And a blue sky spectrum has no yellow in it! [20] Where is the yellow? Something has been absorbed by the sky by scattering or reflection from the clouds. Since Wollaston and Young had maintained that $R + G = Y$, the colour so absorbed is red. The absorption of red, or at least part of the red, degrades the yellow to yellowish-green. But the green or yellowish-green thus produced by absorption in the sky can be rendered yellow again by the interposition of absorbing filters in the beam. [19, p493] This, clearly, will restore the correct intensity ratios for the colours. The degraded green can in like manner also be restored to yellowish-white and even white. One need only choose the correct absorbents. Thus Brewster concluded:

> it follows from Fraunhofer's and Wollaston's and Young's observations taken by themselves, that the quality of the colours of the brightest part of the spectrum is changed from yellow to yellowish-green or green according to Dr Young. And it follows from my observations, combined with theirs (i.e. Fraunhofer, Wollaston and Young) or when standing alone, that in the same part of the spectrum there exist red, yellow, and blue rays of the very same refrangibility. [19, p493]

Once more Brewster had brought a superior argument to refute a detractor.

Regardless of Brewster's success in overcoming objections, there is nevertheless a philosophical objection to his experiments and hence to his theory. Acceptance of experimental results depends upon their being repeatable. Other people must be able to achieve similar results given similar circumstances. All these polemics show that no one was able to duplicate Brewster's results either by their own methods or by their interpretations of his methods. Even Brewster felt some doubt about this for in concluding his reply to Melloni he said, 'I feel assured that my Analysis of the Spectrum will be confirmed by future observers who may repeat my experiments with the same care with which they were made, and without prejudice in favour of their own speculations.' [19, p494] While Brewster was referring rather to experimental methods, his words were prophetic. The person who finally brought repeatability to these experiments was Hermann von Helmholtz four years later. Without apparent prejudice, and with irrefutable experiments, he achieved a new understanding of absorbing media and primary colours. The answers, however, were certainly not what Brewster had in mind in the passage above, for the result was to utterly destroy his theory.

NOTES

1 Herschel, William 1800 Investigation of the powers of the prismatic colours to heat and illuminate objects *Phil. Trans. R. Soc.* **XC** 255

2 Brewster, David 1822 Observations on vision through coloured glasses, and on their application to telescopes, and to microscopes of great magnitude *Edinb. Phil. J.* **VI**

3 Brewster, David 1823 Description of a monochromatic lamp for microscopical purposes, with remarks on the absorption of the prismatic rays by coloured media *Trans. R. Soc. Edinb.* **IX**

4 Herschel, John F W 1823 On the absorption of light by coloured media, and on the colours of the prismatic spectrum exhibited by certain flames *Trans. R. Soc. Edinb.* **IX**

5 Brewster, David 1831 On a new analysis of solar light *Edinb. J. Sci.* **V** also published in 1834 *Trans. R. Soc. Edinb.* **XIII**

6 Newton, Isaac 1671–2 *Phil. Trans. R. Soc.* **80** 3081. Reprinted in I B Cohen (ed) 1958 *Isaac Newton's Papers and Letters on Natural Philosophy* (London: Cambridge University Press) p53

7 Herschel, John F W 1845 Article on light *Encycl. Metrop.* **III**. The article was originally published in 1827. It is sometimes referred to as Herschel's *Treatise on Light*

8 The reference to dyes in the eye is to John Dalton who had proposed this mechanism to explain his own colour blindness. See Chapter 7 for a detailed discussion of colour blindness and Dalton's symptoms

9 Brewster, David 1833 Observations on the absorption of specific rays, in reference to the undulatory theory of light *Phil. Mag.* Ser. 3 **2**

10 Tyndall, John 1882 *Six Lectures on Light* 3rd edn (London) p48

11 Airy, George Biddell 1833 Letter to Sir David Brewster *Phil. Mag.* Ser. 3 **2** 419–24

12 Brewster, David 1837 On the connexion between the phenomena of the absorption of light, and the colours of thin plates *Phil. Trans. R. Soc.* **127**

13 Whewell, William 1847 *History of the Inductive Sciences* 2nd edn (London). There was no commentary by Whewell on this subject in the 1st edition

14 Airy, George Biddell 1847 On Brewster's new analysis of solar light *Phil. Mag.* Ser. 3 **30**

15 Brewster, David 1847 Reply to the Astronomer Royal on the new analysis of solar light *Phil. Mag.* Ser. 3 **30**

16 Draper, John William 1847 On the production of light and heat *Phil. Mag.* Ser. 3 **30** 351–2

17 Brewster, David 1847 Observations on the analysis of the spectrum by absorption *Phil. Mag.* Ser. 3 **30** 462

18 Melloni, Marcedinio 1848 Researches on the radiations of incandescent bodies, and on the elementary colours of the spectrum *Phil. Mag.* Ser. 3 **32**

19 Brewster, David 1848 Observation on the elementary colours of the spectrum, in reply to M Melloni *Phil. Mag.* Ser. 3 **32**

20 This is not correct but would, of course, have been unknown to Brewster. A knowledge of spectral distributions for daylight did not become available until the 1880s. See for example S T Henderson 1970 *Daylight and its Spectrum* (Bristol: Adam Hilger)

3
Hermann von Helmholtz and the Demise of Brewster's Theory

All of Brewster's detractors aimed at showing that he had not seen, nor *could* have seen the observations he claimed to have viewed. Of these essentially there were two, that light changed colour after passing through a filter and the famous 'Brewster's white' light. Airy and Melloni even inferred that their own techniques were superior to his which probably was not so. From each encounter Brewster emerged stronger than ever. He expanded the theory's horizons by explaining the 'cause' of red colour blindness, attempted an explanation of the colours of thin plates in competition with the new wave theory, and he gave the theory a quasi-quantitative basis in his example of isolating spectral orange from red, yellow and blue primaries. As of 1848 nobody it seemed was in a position to overthrow Brewster's concepts, and yet there was deep doubt about their validity for they did not satisfy the demand for repeatability in physical science. Even worse, it was impossible to reconcile Brewster's theory with that of Newton on dispersion and the thought of attempting such a thing was a task no physicist could anticipate with pleasure. In effect Brewster's theory and its nasty, unanswerable questions were a stumbling block to physicists and the only solution seemed to be a complete falsification of his findings. It would be necessary to duplicate Brewster's results exactly, explain all the possible sources of error or misinterpretation and finally perform a critical examination of the hypotheses upon which it rested. The situation was analogous to the question of the phlogiston theory in the 1770s and 1780s. While that theory ostensibly explained many chemical observations it too left lingering doubts such as the necessity of ascribing negative mass to phlogiston under more critical experimental conditions. The overthrow of that theory involved a series of critical experiments, examination of the old hypotheses, and finally a replacement of the old theory with a new one. Just as Lavoisier played a central role in that story, it was Hermann von Helmholtz who played a similar role to Brewster's theory. In this chapter we shall examine Helmholtz's work on both this problem and on the primary colours as described in

Hermann von Helmholtz
(courtesy of Burndy Library).

two pivotal papers of 1852. These papers ultimately set the problem of three-colour vision in an entirely new framework.

If Thomas Young was one of the acutest men who ever lived then certainly Hermann von Helmholtz was another of that ilk. His interest in physics had been aroused while a student at the Potsdam Gymnasium, but not having money to attend university he was advised to study medicine where government support was available. Thus Helmholtz, like Young, began his career as a physician, initially serving as an army surgeon in return for his scholarship aid. However, his brilliance as a research scholar even while in service brought him early recognition so that in 1849 at the age of 28 he was appointed Professor of Physiology at Königsberg. With this he began a lifetime of academic research first in physiology and physiological optics and later in acoustics, pure physics and even mathematics. To each of these fields he made significant contributions. He was the first, for example, to measure the speed of nerve impulses although his old physiology teacher, the renowned Johannes Müller, had delighted in telling everyone that it couldn't be done. In 1847 while still in army service he gave the first clear statement of the principle of conservation of energy, while in 1851 he invented the ophthalmoscope which brought him international fame. Devoting his later career entirely to physics, he made seminal contributions to hydrodynamics and thermodynamics. Among his many writings his treatises on physiological optics and physiological acoustics are great classics of science. If Young was perhaps 'Pentathlos-Beta,' Helmholtz was certainly 'Alpha' for he was a leader in everything he attempted. Helmholtz, therefore, unlike anyone else of his time, was uniquely qualified to view Brewster's experiments from widely different points of view; his overwhelming arguments against Brewster were indeed those of both the physiologist and the physicist. Interestingly, the nineteenth century opening of colour perception as a recognised study was created by a physician–physicist while the next major advance in that field almost 50 years later was also made by a physician–physicist. The close interrelationship between anatomy, optics and aspects of vision almost certainly required this duality of point of view in order to clarify basic concepts.

HELMHOLTZ DEFINES THE PROBLEM

Helmholtz took up the problem in 1852 with a decided advantage. Not only was he a physician–physiologist–physicist but he enjoyed an advantage in time not given to anyone else in this dispute. He could consider historically the debate between Brewster and his antagonists. He read not only Brewster's original papers but the commentaries of Airy, Draper and Melloni together with Brewster's replies. He therefore had, finally, the advantage of knowing Brewster's techniques (mainly from his replies) and so could avoid the 'heinous' error of not exactly repeating Brewster's experiments. He moreover took the charitable point of view that Brewster had indeed seen what he claimed to see in the spectrum and set out to accomplish essentially three main goals: (i) repeat the experiments according to Brewster's methods in order to achieve Brewster's results, (ii) analyse all the optical and *physiological* errors inherent in the experiments leading to such results, and (iii) repeat the experiments with the errors removed to see whether new results appeared or whether Brewster's results remained. The implicit purpose, of course, was to completely falsify the whole theory.

One point made at the outset by Helmholtz refers to Brewster's choice of primary spectral colours. Helmholtz stated that the red, yellow and blue primaries chosen by Brewster, although the colours usually chosen for such theories at that time, were not correct choices; for the choices had been based upon mixtures of pigments rather than of light. [1, p402] For example, using these particular colours one assumed that yellow and blue make green. But Helmholtz stated unequivocally that yellow and blue lights do not make green but white. [1, p402] Therefore, said Helmholtz, we cannot retain these colours as our choice of primary colours but must have others, perhaps red, green and violet as advocated by Thomas Young. However, the choice of primary colours does not really affect Brewster's theory. Except for changing the detail, the essence of the theory that spectral colours are modified by absorption remains unchanged. Thus, Helmholtz passed to the problem as to whether homogeneous light is or is not altered by passing through coloured media.

Helmholtz repeated Brewster's experiments using a smalt-blue glass similar to Brewster's with the object of seeing what Brewster had seen. However, mindful of the necessity of piecing together Brewster's work mostly from his replies to criticism, Helmholtz asked Brewster's indulgence 'in case I should allude to possible sources of error which he has taken pains to avoid.' [1, p403] The results were gratifying. The yellow part of a daylight spectrum was quite feeble when viewed through a single glass plate, but the spectrum of the sky near but not directly looking at the sun showed a 'pure and shining yellow'. [1, p408] With two plates of smalt glass the feeble yellow of the daylight spectrum disappeared while in the near-direct spectrum yellow appeared almost white. Violet and extreme red continued to be seen brightly through the two plates. Through three plates the yellow band seemed bluish-white. In accordance with Brewster's request for repeatability Helmholtz had actually seen 'Brewster's white' and the apparent changes in colour which Brewster had ascribed to the absorbing media. Of this Helmholtz said:

> A careful repetition of at least the most important of his experiments, carried out in exact accordance with his method, and with every precaution hitherto deemed necessary, has indeed taught me that the facts which he affirms to have observed are described with perfect accuracy; indeed nothing else could be expected from so skillful an observer. [1, p403]

Moreover, Helmholtz claimed that there was no overlapping of colours as

Melloni had thought. On the contrary, he recognised the great merit of Brewster's later experiments on Fraunhofer lines in which he had counted over 2000 lines and so it appeared to him that Brewster had possessed better apparatus for colour separation by refraction than had Fraunhofer himself. Thus, in short, Helmholtz set aside the judgements of Brewster's critics and justified Brewster's observations. The problem now was to explain why these phenomena appeared.

The first question was: What was the source of the 'homogeneous white light' Brewster (and he) had seen? Was there not a source of dispersed white light possibly mixed in with thc spectrum? The spectral colours were undoubtedly separated completely so no overlapping took place. Also the amount of light entering the apparatus could be carefully limited by the aperture or slit so that problems of extraneous light, one of Airy's main objections, were eliminated. Therefore, any dispersed light besides that of the spectrum had to be sought for within the apparatus itself, a new and novel conception of the problem, and from Helmholtz's point of view, at least part of the apparatus was the eye itself.

SOURCES OF ADDITIONAL DISPERSED LIGHT

The prism itself was an important source of additional dispersed white light. No matter how clear the glass nor how highly polished the refracting surfaces, one could always see in the prism interior many shining points and minute scratches. [1, p404] Internal irregularities and tiny surface scratches create dispersed light in an irregular pattern. As this is a sensible quantity it gives the prism a rather 'smoky appearance.' [1, p404] Such an appearance can be easily demonstrated according to Helmholtz by allowing a pencil of light to fall upon the prism. If the eye then intercepts the transmitted rays in such a way that the transmitted light does not actually enter the eye, 'the little irregularities of the surface and mass then appear brightly illuminated against the black ground furnished by the screen.' [1, p404].

Another prismatic problem arises from repeated internal reflections from its surfaces. Helmholtz referred to a five-sided prism being used in these experiments, something Brewster certainly never mentioned. In these prisms two sides only are polished and hence used for refraction, the other three sides being ground dull. The internal reflections of such prisms can be demonstrated by placing the prism against a dark background and illuminating one of the dull surfaces. One can then see within the prism a series of reflected images of the dull surface [1, p405] since the two polished sides act as angular mirrors giving multiple reflections of the dull side. As used in these experiments, the observer looks through one of the polished sides into the prism which is tantamount to looking through one of the reflectors. Then the reflected image of the dull side is seen in exactly the same direction as the spectrum one observes normally. This is not necessarily serious if no incident light falls on the dull surface, but Helmholtz stated without hesitation that some of the light *does* normally fall on the dull side. This means that as the observer looks through (or into) the prism he will see not only the spectrum as expected but also a reflection of the dull side superimposed over the spectrum. The result is a weak light luminosity spread over the spectrum. Now this effect is certainly small and probably not usually observed beside the regularly refracted light, but we can easily see that this together with the irregular dispersion previously mentioned could possibly lead to Brewster's 'homogeneous' white light. Helmholtz recommended that the three dull sides of the prism should be blackened to eliminate this effect. [1, p405]

There is a third possible optical problem with Brewster's experiments involving the actual absorbing media. The glass filters themselves must be considered as small but significant causes of dispersion for at least two reasons. [1, p405] Firstly, the glass itself was not a uniform medium and there would therefore have been internal dispersion by mass points within the glass filter as in the case of the prism. Secondly, we cannot be sure the opposing surfaces were exactly plane and parallel. In fact the glass specimens used in these types of experiments were basically any bits and pieces of glass, mainly commercially manufactured for windows, to hand. They certainly could not be expected to have plane-parallel surfaces or uniform density. Therefore, the non-parallel sides would produce an image in the usual manner of lenses and light reflected internally from the second surface to the first surface and back to the second would enable the second surface to form its own image.

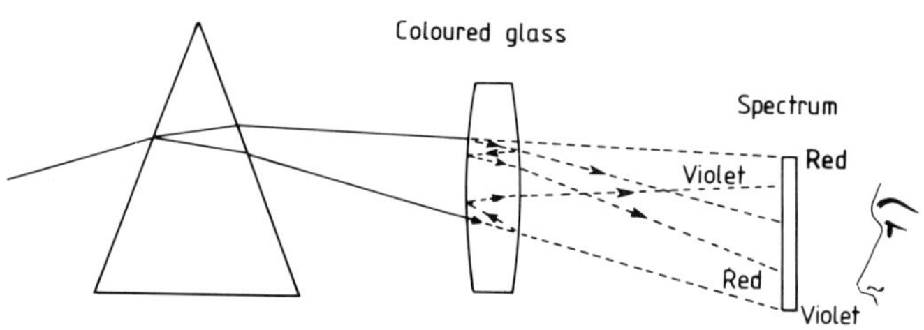

Figure 3.1 Multiple internal reflections in a glass absorber as envisioned by Helmholtz. The curvature of the glass is exaggerated.

Essentially the filter acted as two additional lenses in front of the prism or eye whose secondary images fell upon the colours of the primary image produced by the prism (see Figure 3.1). Although Helmholtz did not mention it, it seems possible that besides seeing colours somewhat displaced from their usual positions there was some colour mixing here from the secondary images of the filters with the primary image of the prism. Thus if some red light from the secondary image fell on the green of the primary image there would be a shift of the character of the spectrum to the yellow and this would account partially, at least, for the spreading of the yellow into the red and green spaces which had so much impressed Brewster. We shall see presently, however, that there is another explanation for this effect dealing with colour sensation as a function of contrast.

The filters as used by Brewster contribute still another problem; and it is here that Helmholtz's training as a physiologist and his interest in the eye itself become evident. Light emerging from the filters finally enters the eye, but the first optical surface of the eye is the cornea which reflects some of the incident light. This light must be reflected back to the filter surface and from there back again to the eye. As Helmholtz stated, 'the image thus formed being too near the eye must appear as a bright luminosity in the field of view.' [1, p405] All these filtering problems can be eliminated or much reduced by the simple expediency of placing the filters between the light source and aperture instead of between the prism and eye, as Brewster had done. [1, p406] Airy, we recall, had followed this procedure in his experiments of 1833 (for convenience sake) which lends some credibility to his stated results.

We have now seen several separate mechanisms for producing weak light luminosities distributed over the primary spectra. Probably none of these effects singly are significant but collectively they could contribute a noticeable effect, especially 'when added to colours that have already been greatly weakened by absorption.' [1, p406] These then could readily account for Brewster's homogeneous white light or at least produce subtle changes of hue leading to false observations.

However, considering Brewster's replies to Airy and Melloni, Helmholtz thought it just possible that Brewster might have eliminated these sources of error, which if so was never made clear by Brewster. Nevertheless, giving him the benefit of the doubt, Helmholtz looked for and found other effects within the eye itself which Brewster could not possibly have eliminated. Here again Helmholtz as a physiologist brought a new understanding to observations of colour not previously considered by physicists thinking in optical terms only. We have, in fact, already had a hint of this new point of view in Helmholtz's mentioning reflection from the cornea to the glass absorbers.

PHYSIOLOGICAL ERRORS

The weak light luminosity giving the illusion of 'Brewster's white' can be attributed to at least two effects of the eye. This first was mentioned by Helmholtz who stated that:

> when very bright light of any kind whatever falls upon a portion of the retina, light of the same kind appears diffused as a weak luminosity over a great portion of the field of view. [1, p406]

For example, Helmholtz described an easy way of observing it by viewing a candle in a darkened room against a dark background. Using one eye only, look at the dark background by blocking the light from the candle to the eye with a finger. Then remove the finger to see the candle itself. Alternating the observation of the dark background and the direct candle light,

> it will be readily seen, that as often as the rays freely enter the eye a white luminosity appears spread over the surface, being brighter in the

vicinity of the light, and spreading itself weakly over the more distant portions of the surface. The same is observed when daylight, and most strikingly when direct sunlight, enters the eye from an orifice in a dark screen. [1, p406]

The question immediately arises as to the cause of this effect. Ruling out possible subjective reasons within the nerve system of the retina, Helmholtz argued that dispersion within the eye itself causes light to reach 'other portions of the retina than those affected by the regularly refracted light.' [1, pp406–7] Diffraction by the pupil, for instance, although quite small, he thought was probably the main cause. If the distance from the pupil to the retina were sufficiently large we should get diffraction rings on the retina and the separation of colours this implies. There would be alternating circles of bright and dark on the retina with the bright circles comprised of different colours. But in fact the retina is close to the pupil so the areas of the various rings might well overlap giving another source of weak light intensity. Therefore Helmholtz argued that the pupil of the eye itself in conjunction with its optically short anatomical distance from the retina must be considered the source of the weak luminosity referred to above. (This, however, is probably not correct as the diameters of the pupil's diffraction rings are very small, their areas are also very small and the proposed overlapping cannot really account for the weak light diffusion.)

Another anatomical factor contributes to the effect. In 1851 Helmholtz had invented the ophthalmoscope, by which for the first time one could observe the retinas of living subjects. With this he found that a noticeable quantity of light from the illuminated portion of the retina was reflected from the retina back to the pupil. As in any refracting system, this light would then be partially transmitted through the lens system but partially reflected back again. Therefore, Helmholtz felt that the light reflecting *from* the retina would be reflected back *to* the retina from the cornea. He ventured the opinion that 'it is therefore to my mind an undoubted fact, that a portion of the light incident upon the eye is deflected so as to fall upon other portions of the retina.' [1, p407]

Helmholtz was now in a position to understand Brewster's most unusual result, the one of which he was most proud. That is his 'isolation' of homogeneous white light in the yellow part of the spectrum as described in the analysis of the spectrum in 1831. In another paper Helmholtz had shown the results of mixing colours of the spectrum. [2] The result was that mixtures of indigo-blue and yellow in different proportions produced a whitish-yellow, then white and finally a bluish-white. Now if some of the dispersed light, for all the reasons mentioned, fell on the retina in the same place as the yellow which passed through the absorbent, then the colours observed would certainly not have been exactly yellow. Referring to the smalt-glass observations Helmholtz said:

> the colours in the smalt-glass spectrum which lie next to the yellow, namely red and green, can, by mixing in various proportions, cause the

> white to approximate to red or green, as the case may be, and thus exhibit all the degrees of colour which are observed through different thicknesses of the blue glass. [1, p408–9]

The discussion above explains Brewster's observation of white in the yellow part of the spectrum. It also gives plausible reasons for colour changes when absorbents are placed in front of the eye. It is, in fact, due to both optical dispersion and a physiological effect of producing compound colours on the retina of the eye as a result of the dispersion.

To explain the origins of Brewster's observations was certainly an important contribution towards toppling his theory. But alone it was not enough. Such had been the approach after all of Airy, Draper and Melloni with no acceptance of their results. Even allowing for the novel, perceptive arguments of Helmholtz and giving Brewster all benefit of the doubt these were still not definitive arguments against Brewster's theory. They constituted, in the end, only plausible hypotheses. What was really needed was a means of testing the hypotheses themselves. In other words, Helmholtz's concept of colour mixing on the retina itself needed to be tested by repeating the absorption experiments in which 'the disturbing colours shall be totally, or almost totally, excluded from the field of view.' [1, p409]

BREWSTER'S EXPERIMENTS WITH THE ERRORS REMOVED

It was important to view the colours directly as they emerged from the prism in order to follow Brewster's procedure as closely as possible. Airy, especially, had caused Brewster considerable indignation by projecting the colours on a screen. Helmholtz hit upon a happy modification of Brewster's method which he described as follows:

> Solar rays reflected from a mirror are permitted to enter through a narrow slit into a dark room and to fall upon a vertical prism. Immediately behind the latter is a lens which casts the spectrum formed by the prism upon a screen. In the latter is a second very fine vertical slit. The light of that band of the spectrum which falls exactly upon the slit passes through, while the rest is cut off. The observer stands behind this second screen, the back of which is well blackened, best covered with black velvet, and looks at the slit through a second prism of the best possible quality. [1, p409]

This method bears a marked resemblance to an experiment by Newton, which he called his 'Experimentum Crucis.' [3, p3078] Suspecting after several experiments that the different spectral colours exhibited unequal angles of refraction (for a given angle of incidence), Newton set out to confirm his suspicions. We can see the similarity from Newton's diagram of his experiment shown in Figure 3.2. [4] Sunlight S enters the room at F, passes through the prism ABC, and the resulting spectrum is incident at

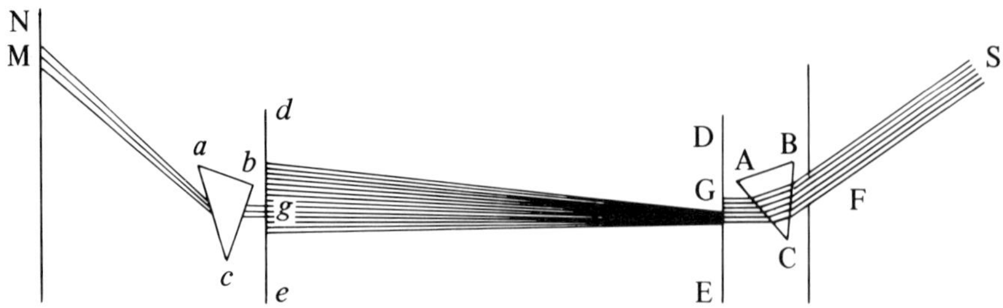

Figure 3.2 Newton's 'Experimentum Crucis' as given in the *Opticks*. [4] Sunlight in the beam is dispersed by the prism ABC but a narrow portion of the resulting beam shows no additional dispersion by the second prism abc. Separation of the colours takes place when different colours of light exhibit unequal angles of refraction.

the hole G in a board DE. The light passes through to another hole *g* in a second board *de* and then through a second prism *abc* for further analysis. By rotating the prism ABC, different parts of the spectrum were incident at *g*, and the different colours reached different places on the screen between M and N. Since the angle of incidence on the second prism was the same for all colours because of the collimating effect of the two holes, G and *g*, Newton had his proof that 'Light consists of Rays differently refrangible.' [3, p3079]

Helmholtz's method was essentially the same. The narrow slit admitting light in his arrangement served the dual function of Newton's openings F and G. He then had a second slit to select only a portion of the spectrum for analysis and a second prism to perform that analysis. Helmholtz's method differed from that of Newton's basically by the lens which improved the spectral image. In fact, Newton in other experiments, also used a lens for the same reason. With the close similarity of his arrangement to that of Newton's, Helmholtz in effect returned to fundamentals. He set out to determine once again whether or not light of different colours is differently refracted. It is a striking fact that as a result of Brewster's theory it was ultimately necessary to reconfirm Newton's most fundamental optical discovery some 180 years later!

Let us return to Helmholtz's experiment. If there were no irregular dispersion in the first prism (or in the lens) only homogeneous light of a particular colour would be incident on the second slit. After passing through the second slit and the second prism, there would not be a spectrum but only a narrow band of colour. But there is irregular dispersion as mentioned before due to surface scratches, internal reflections and non-homogeneities in the glass. Therefore, some of this white light will enter the second slit. The result is that this light will be regularly dispersed by the second prism thus forming a weakly luminous spectrum superimposed on the bright band of colour selected from the primary spectrum formed by the first prism. But this is allowable since dispersion by the second prism simply superimposes on the primary band of colour (from the first prism) the same colour from the irregular white. Problems of irregular dispersion due to the second prism are below the level of sensibility of the observer. [1, p409] This eliminates colour mixing on the retina of portions of the irregularly dispersed white light with portions of the regularly refracted spectrum.

Now came the telling observation against Brewster. Helmholtz used this method to isolate the yellow band produced by the smalt-glass. This he then tested by introducing additional plates of smalt glass before either of the slits. The results were now not the same as Brewster's although the method itself was essentially his. Where Brewster had described a change in the yellow produced by additional plates of glass and where Helmholtz had confirmed this without the modification above, now no change of colour was observed. As Helmholtz said, 'the yellow retains its originally pure and saturated colour after it has passed through two, three, or even four plates of the blue glass.' [1, p410]

Recall here that probably the most important observation Brewster made was with the blue glass. Through this glass he had seen a change in spectral yellow together with its spreading out beyond its usual boundaries. From this came his hypothesis of different colours of light possessing equal indices of refraction. If Brewster were correct, the change of hue of yellow together with the spreading effect should still have taken place. On the other hand if Helmholtz's hypothesis about colour mixing in the eye were correct we should expect no change in the yellow light with additional filters since the other colours have been eliminated. The observation above clearly confirmed the Helmholtz prediction and Brewster's theory became irreconcilable with this observation. [1, p410] Helmholtz stated that:

> the light of the yellow band on entering the first slit appears actually whitish; but when it is viewed through a second prism, it is decomposed into pure yellow and light of other colours; hence it does not possess the same refrangibility, but, in accordance with the explanation given by me, is a mixture of rays of different refrangibilities. In Brewster's proceeding, a mixture of foreign light, whether in the prism, glass plates or eye could not be avoided. From this point of view it is quite intelligible how the introduction of coloured media might render the white colour of the band in question more pure, or cause it to approximate to red and green. [1, p410].

PSYCHOLOGICAL EFFECTS IN COLOUR PERCEPTION

The optical analysis and experiments already performed by Helmholtz would have been sufficient to undermine Brewster's theory. But there were still other sources of error involving psychological problems in colour interpretation as opposed to purely optical or physiological phenomena. Helmholtz was much impressed with subjective effects and colour illusions due to contrast and relative brightness of colours, a subject then being much investigated by people such as Ernst Wilhelm Brücke, Jan Evangelista Purkinje, Helmholtz himself and others. These investigations were part of a general interest in optical illusions, which, ironically, originated in a letter from L A Necker to Brewster in 1832 describing his now famous transparent reversing rhomboid. [5]

Brücke had in 1852 pointed out that the appearance of certain colours depends upon the brightness of adjacent colours. [6] Contrast effects had also been investigated by the French chemist, Eugene Chevreul (see Chapter 4). Repeating Brücke's experiments, Helmholtz found specific contrast effects as follows: (i) bright light next to dark portions of the field of view stimulates the same colour in the dark field as the bright light; (ii) weak light adjacent to dark portions stimulates complementary colour in the darker section; (iii) the complementary colour stimulated is brighter when the eye moves as opposed to fixing on a point; and (iv) the complementary of red is more easily induced than the complementary of green. [1, p411]

These results were most interesting in terms of yet another of Brewster's remarkable observations. He claimed to have seen green light in the red part of the spectrum. Said Brewster:

> When the most refrangible rays are absorbed by certain thicknesses of balsam of sulphur, balsam of Peru, pitch, or red mica, the blue mixed with yellow, and forming green, may be traced very near the C line of FRAUNHOFER, which is considerably within the red space. That the blue extends over the whole red space is proved by the same arguments which we used for yellow light; for when the red space is made of an orange tinge by the absorbtive action of certain yellow, orange, and green media, this change can be effected only by the absorption of blue light. [7]

Brewster's homogeneous white can be explained by irregular dispersion and his claim of yellow spreading into the red and green is explained by colour mixing in the eye. But this observation cannot be explained by either hypothesis. It is, in fact, an illusion caused by the effects of contrast in the eye.

Repeating Brewster's experiments, Helmholtz used as absorbing media Peruvian balsam, sulphur-balsam and pitch (all brown liquids). [1, p411] With thin layers of the absorber he saw a spectrum with clear red, yellow and green but no blue or violet. In the absence of the other colours green appeared to extend into the yellow and even into the reddish-orange immediately adjacent to the red, while the yellow and orange parts of the spectrum also appeared green. [1, p411] The effect was 'so vivid that it is indeed difficult to conceive that it could be a subjective illusion.' [1, p411] That it was an illusion however, was easily demonstrated using the previous method of isolating colours. Then the colours observed through the brown liquids appeared completely unchanged and exhibited no tendency toward green. [1, p412]

What then was the explanation for the appearance of green next to the red? Helmholtz first remarked that the extent of the green was greater when the eye wandered over the spectrum than when it was fixed on the green portion of the spectrum. [1, p412] The effect of moving the eye from the green towards the red was to allow yellow light to strike the retina where it had formerly viewed the green. This tended to generate a blue-green secondary colour. Moreover, weak orange light (produced by a thicker layer of the absorbents) next to a bright red can appear green. This could be easily demonstrated by

> sticking a small disc of paper coloured red by vermillion upon a plate of red glass, and holding the latter against a very bright ground, the bright firmament for example, while the disc is only weakly illuminated. With a suitable strength of illumination it appears green. [1, p412] [8]

These experiments on contrast and induced complementary colours indicated that subjective effects could take place in observing the colours of the spectrum and that these effects could explain some of Brewster's observations. Airy was right in denying memory for colour to the eye and Brewster was wrong in supposing, contrarily, that his eye after many years practice did possess a colour memory. Not only does it not have a colour memory but it is easily fooled. In terms of subjective changes of colour Airy's method of spectral comparison on a screen (learned from Brewster) was certainly the better method. By doing so he had (unwittingly) eliminated or reduced subjective effects while reducing irregular dispersion and colour mixing problems by placing his absorbing media before the prism rather than after it. Therefore, he achieved nothing like Brewster's results. Considering Helmholtz's work, Airy's critique of Brewster in 1847 becomes more meaningful and even tends to corroborate some of these results. If Brewster thought of this possibility after seeing Helmholtz's paper, he did not acknowledge it.

The contrast phenomenon also explains an observation by John Herschel where in viewing a spectrum through a smalt glass the transition from red to yellow was quite sudden 'and the contrast of colours quite striking.' [9] (It was a fortuitous choice of words.) What then, Herschel asked, had become of the orange? He thought it possible that a colour change had taken place as a result of absorption and that Brewster's theory might just have some real observational basis. Brewster had invoked this observation in his reply to Airy. However, this too must have been a colour illusion. Although inexplicable in terms of induced secondary colours, the apparent juxtaposition of the red and yellow was still an illusion giving the impression that the yellow had spread over the orange portion of the spectrum. Isolating one band of colour from the other Helmholtz showed that all the spectral colours were in their proper positions and therefore no colour changes took place as a result of absorption. Newton was right after all.

Finally, another physiological phenomenon subtly alters the sensation of colour. Colour sensation in the sense of tint (or saturation) as distinguished from wavelength or hue depends on brightness, so that the same wavelength at different intensities actually gives different impressions. [1, p411] At very high luminosities all colours appear white, a fact hinted at as early as 1704 by Newton, [4] J H Lambert in 1772, [10] and most closely in time to Helmholtz by J D Forbes in 1849. [11, p161] Indeed Forbes had said that the brightness of coloured lights projected on a screen resulted from the sum of their luminous intensities 'and if a sufficient number of rays be combined, the result will be a dazzling white.' [11, p165] However, the clear concept of colour sensation as a function of brightness (as well as wavelength) escaped all these investigators and was only arrived at in 1853 by the mathematician Hermann Günther Grassmann and Helmholtz himself. Applying this idea to an absorbing solution of a copper–ammonium salt, Helmholtz stated:

> if, therefore, a certain thickness of the solution of ammonio-sulphate of copper shows the blue of the spectrum bright and whitish, while a greater thickness causes it to appear a deep dark blue, we are simply to conclude that this fluid absorbs blue rays, but by no means that it has abstracted white light from the blue. [1, p414]

We now finally come to an understanding of the four colour spectrum obtained by Young and Wollaston, which had shown only red, green, blue and violet. According to Brewster it was a daylight spectrum, not one of direct sunlight. Brewster had explained the absence of yellow and orange in the spectrum as due to absorption from scattering and cloud reflection. Yellow was scattered or absorbed and since orange was a combination of red plus yellow there could also be no orange. This, of course, is wrong and the explanation lies in varying intensities of colours in the spectrum either of daylight or direct sunlight. The observation to explain is simply this: yellow in the solar spectrum occupies a prominent wide space while in the daylight spectrum it is much less discernible.

The first point noted by Helmholtz was that not only is yellow in the solar spectrum prominent, it is the *most* prominent colour appearing with 'a dazzling brilliancy.' [1, p414] Also green and red in the solar spectrum have considerable intensity and can appear yellowish if bright enough. Therefore, yellow appears most prominent. On the other hand, all the colours of the daylight spectrum are attenuated so that the apparent change of colour of red and green into yellow does not take place. The yellow which is ostensibly absent in such a spectrum can definitely be seen either by viewing the spectrum under large magnification or by the isolation method already used so successfully. Helmholtz confirmed this by converting, in effect, a direct spectrum into a daylight spectrum. Viewing the direct spectrum through a pair of crossed Nichol prisms, the solar spectrum could be so attentuated that the yellow diminished and the spectrum as a whole took on the appearance of a daylight spectrum. [1, p415] The same effect could be produced by several reflections of the spectrum from uncoated glass plates which would reduce the overall brightness but not necessarily equally for all colours.

It appears from this that the explanations by Herschel and Brewster in 1822 of Wollaston's observation were incorrect. Using absorbing media they attempted to show that yellow did have an independent existence in the spectrum as contrasted to its being a combination of red and green as Young thought. They succeeded in doing so but by using apparently a direct spectrum instead of a daylight spectrum as Wollaston had done. Although Brewster later pointed out that Wollaston's spectrum was a daylight spectrum, in his own original work of 1823 he mentioned forming 'a very brilliant spectrum' [12] and then adjusting the position of the prism to duplicate Wollaston's spectrum. Apparently Brewster used a direct solar spectrum here. And it was from this attempt on the part of Brewster that the main observations leading to his 'New Analysis of Solar Light' arose. How extraordinary it is that a curious and simple observation in 1802 led to some unusual events: for Young it led to a correct choice of primary colours of light for visual purposes; for Brewster, it led to an important contribution to his theory of the triple spectrum; and this, in turn, stimulated some remarkable results in the case of Helmholtz.

CONCLUDING REMARKS ON BREWSTER

Brewster's theory ultimately proved completely false and sterile: but superficially it seemed so strong as to block further progress in understanding colour perception, at least in terms of the composition of the spectrum. Thus, it was necessary to overthrow Brewster's theory. While theory falsification is almost always accompanied by arguments for acceptance of an alternative, this was not required here. The alternative already existed in the form of Newton's classic ideas by then transposed to the undulatory framework. Brewster's theory was simply shown to be based on completely false observations arising from both optical and subjective causes. This pivotal work of Helmholtz was therefore not so much a contribution of new knowledge to the problem of colour perception as a necessary sweeping away of an obstacle. Helmholtz himself was aware of his destructive role. He stated that his own work

> abundantly proves that in his (i.e. Brewster's) method many hitherto unobserved influences come into play, which render a sure judgement of the colours impossible and deprive his arguments of all force. If the assumed connexion of the refrangibility or length of wave with colour is to be proved erroneous, it must be done by some more certain method of observation, similar, for example, to that which I have described in this memoir; a principal condition of which is, that the colour investigated be separated from the other colours and rendered free from every trace of irregularly dispersed light. [1, p416]

After all this it is amusing yet rather pathetic to find Brewster still defending his theory three years later in 1855. In his biography of Newton, Brewster saw fit to expound his own theory in contradiction to the Newtonian system and to answer Helmholtz's criticisms. His answer was a far cry from the ringing, indignant blasts aimed at Airy and Melloni. As to the influences due to irregular reflection and colour mixing, Brewster allowed that it would be influential and that was the extent of his answer. As for subjective complementary colours he could only answer 'if it does disturb the colourific impressions on retinae tender and sensitive, (it) had no such effect on ours.' [13, p124] Defending himself to the last, Brewster could only say that if these subjective effects were really so, then all the hues of nature must be false:

> We must not now pronounce a rose to be red, and its leaves green, till we have stared at them through a chink, or torn them from their foot-stalk! The phenomena of absorption which we have described *we have seen*, just as Newton saw his seven colours in the spectrum, and Hooke his composite tints in the soap-bubble; and now that our eyes have nearly finished their work, we are not disposed to mistrust, without reason, such good and faithful servants. [13, pp124–5]

He hardly seemed to realise that Helmholtz had clearly stated that he too had seen Brewster's results under the same conditions. He simply rejected out of hand all other possible reasons for his optical observations implying that his were perfectly straightforward and not subject to any subtleties.

Finally at the meeting of the British Association for 1855, Brewster repeated these remarks while adding several others in the same vein. None of them was very convincing and none of them was answered. Brewer himself retired from the battle, unvanquished, with a scathing conclusion:

> I may now leave it to my friend Professor Wartmann, of Geneva, our greatest authority on the subject of colour blindness, to determine which of all the parties in this controversey are afflicted with that optical disease. [14]

Obviously he didn't think he was so afflicted! But these last remarks make a hollow defence after all. They are the arguments of a literary person rather than an objective scientist and in this respect we may note, as earlier, that Brewster supported himself by his considerable literary efforts rather than by science. Since he was then 64 years old it must certainly have been a painful experience to see one of his finest 'creations', perhaps the best of all in his own estimation, taken to pieces and turned to ashes in his mouth. It may have been particularly hard for him since the work of all his detractors was published in the *Philosophical Magazine* of which he was editor. In selecting Helmholtz's great paper (and another as well) for publication in his journal he rather monitored his own decline. In stubbornly maintaining his own theory against the work of Helmholtz and that of Grassmann and Maxwell which shortly followed, Brewster rejected all current advances in a field he himself had pioneered. He lived long enough—to 1868—to see himself surpassed in his own field of optics.

Unfortunately the subject of colour perception has perforce dwelled on one of Brewster's errors without having the opportunity of mentioning his successes. He was really a better physicist than this narrative indicates. As mentioned before, he was a leading scientist of his time, made many important contributions to optics and deservedly enjoyed a high reputation among his peers in an age when science reached a new maturity. Combined with this was a voluminous literary output by which he greatly helped to popularise science. And finally he played a leading role in organising the British Association, in obtaining support and recognition for scientific work from government and in reforming Scottish education. In short he led a long and full life and Richard S Westfall has said, he 'stamped his image prominently on his age.' [15] Therefore, let us honour his spirit for his many important contributions to society at large.

NOTES

1 von Helmholtz, Hermann 1852 On Sir David Brewster's new analysis of solar light *Ann. Phys., Lpz* **8.** (English translation 1852 *Phil. Mag.* **IV.** Brewster was editor of *Phil. Mag.* and the translator appears to have been John Tyndall.) References are to the English version

2 von Helmholtz, Hermann 1852 On the theory of compound colours *Ann. Phys., Lpz.* **9** 519 (English translation 1852 *Phil. Mag.* **IV**). See Chapter 5 for a detailed discussion of this important paper

3 Newton, Isaac 1671–2 *Phil. Trans. R. Soc.* **80**. Reprinted in I B Cohen (ed) 1958 *Isaac Newton's Papers and Letters on Natural Philosophy* (London: Cambridge University Press)

4 Newton, Isaac 1704 *Opticks* Book 1, Plate 4 (London)

5 Gregory, Richard L 1968 Visual illusions *Sci. Am.* November 66

6 Brücke, Ernst Wilhelm 1852 Untersuchungen uber subjektiv Farben *Denkschr. Akad. Wiss., Wien, Mat.–nat. Kl.* **III** 95–108 also published in 1952 *Archs. Sci. Phys. Nat.* **XIX** 121–35

7 Brewster, David 1834 On a new analysis of solar light *Trans. R. Soc. Edinb.* **XII** 132–3

8 The translator, John Tyndall, added the following note: 'The subjective colour is very strikingly exhibited in this experiment. A red wafer answers the purpose perfectly; the observer stands in a weakly illuminated place and looks at the sky, the wafer appears a vivid green or blue-green'

9 Herschel, John F W 1845 Article on light *Encycl. Metrop.* **III** 494

10 Lambert, Johann Heinrich 1772 *Beshreibung einer Ausgemahlten Farben-pyramide* (Berlin)

11 Forbes, J D 1849 Hints towards a classification of colours *Phil. Mag.* **34**

12 Brewster, David 1823 Description of a monochromatic lamp for microscopical purposes, with remarks on the absorption of the prismatic rays by coloured media. *Trans. R. Soc. Edinb.* **IX** 411

13 Brewster, David 1855 *Memoirs of Sir Isaac Newton* (Edinburgh). Reprinted 1965 Vol I (New York and London: Johnson Reprint Corp.)

14 Report of the meeting of the British Association for 1855 in *The Athanaeum,* 1855 p1157. The official British Association report is much shorter than this account

15 Westfall, Richard S 1965 Introduction to Brewster's *Memoirs of Sir Isaac Newton* [13]

4
The Organisation of Colour: The Primary Colours

A tradition of philosophical and artistic enquiry into the nature of colour can be traced from antiquity to the present. [1] Classically it has been the business of artists to imitate the colours of nature which they must do with a limited number of pigments. As a practical necessity, quite aside from scientific enquiry, the question of primary and secondary colours arises. Pliny mentioned that the early Greek painters used four basic colours: white, yellow, red and black and such simple palettes seem to have been the heritage of art until the Renaissance. [2] But Leonardo da Vinci apparently used six basic colours: white, yellow, green, blue, red and black considering yellow, blue, green and red as primaries. In both traditions, the arts and philosophy, it was always simply assumed that primary colours existed with an objective reality. Then the problem was to decide how many primaries there really were and how other hues were formed from them. (Did those intrepid artists of the caves of Lasceaux also ponder such problems?) We have already seen that this theme greatly influenced events in the early nineteenth century; and it was to exert a more profound influence on colour science later. We shall trace this history from the seventeenth century to Helmholtz in 1852 when a clear understanding of primary colours finally emerged. Suffice it to say for the earlier tradition that little progress was made in understanding primary colours. As with other modern sciences, a new beginning came in the seventeenth century as part of the intellectual revolution of the time.

Colour has always been one of the most natural of phenomena to investigate and also one of the most difficult. Not being readily amenable to quantitative techniques, early studies of colour centred on colour classification schemes based on assumed primary colours. In the modern era one of the earliest formal attempts at organising colour schemes was made in 1686 by Richard Waller, Secretary of the Royal Society. [3, p24] Waller's purpose was to give a useful table of colours for limning and painting and he confidently expected the table to 'be of some use and advantage in describing the Colours of Natural Bodies.' [3, p25] A table appended to his paper determined the composition of each of Waller's colours, most of them binary combinations of only red, yellow and blue. Said Waller, 'I have mixt each of the Simple Yellows and Reds with each of the Simple Blews, and these Mixtures give most of the mean colours, viz. Greens, Purples, etc.' [3, p24] Aside from this early formal attempt at classifying colours, Waller's terminology is notable for it indicates that red, yellow and blue were already then rather generally accepted as primary colours. In fact even Robert Boyle had said as much twenty-four years earlier, noting that these colours together with white and black were those used by painters. [4]

Newton's colour circle in the *Opticks* given as an empirical rule for compounding colours, is also a colour classification scheme. Because of its importance to the rest of this history it will be described here in detail. The circle is given by Newton as a means of answering the following question:

> In a mixture of Primary Colours, the Quantity and Quality of each being given, to know the Colour of the Compound. [5, p154]

To find the proper proportions Newton imagined a circle with its circumference apportioned to the various colours (see Figure 4.1). The colours are to be homogeneous spectral colours with a gradual passing of one colour into the next. The circle is, in effect, a rectangular spectrum bent into a cylinder and then viewed in cross section. It is not clear whether the circle was to be divided in the same proportions as the colours occupied in the spectrum for Newton, but the inference is clear that the seven spectral colours were here considered as primary colours. Indeed, a choice of so many primary colours seems to be in the tradition of the arts and from this point of view then not very surprising. However, Newton enumerated seven colours in the spectrum rather than six (or five) by inserting indigo between blue and violet in order to effect an analogy between the spectrum and the seven notes of the musical scale. Significantly red and violet adjoin each other allowing purple, which is not a spectral colour, to be produced.

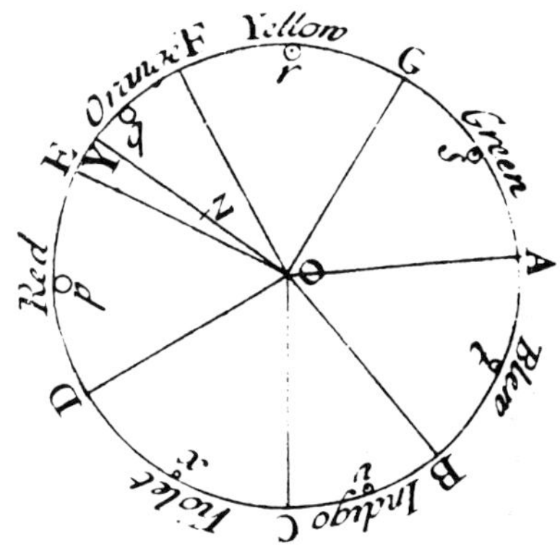

Figure 4.1 Newton's colour circle from the *Opticks*. [5]

In the section from F to G there are all degrees of yellow and G represents the mean colour between yellow and green. In the section from G to A are all degrees of green and A represents the mean colour between green and blue, etc. The tiny circle, r, represents the centre of gravity of the arc FG. The other circles, p, q, s, etc, represent the centres of gravity of the arcs DE, EF, GA, etc respectively. The area of each circle is set proportional 'to the number of Rays of each Colour in the Given Mixture to be described,' [5, p155] or the brightness of the colour. To find the result of mixing all the colours the procedure is: (i) Find the common centre of gravity of all the circles p, q, r, s, etc. This is represented for the hypothetical case at Z. (ii) Draw a line from the centre of the circle, O, through Z to the circumference. We have then the line OY on the diagram. The actual colour is, of course, at Z but the point Y on the circumference shows the part of the spectrum to which this hue belongs; the segment OZ is 'proportional to the Fulness or Intenseness of the Colour, that is, to its distance from whiteneness.' [5, p156] Note that the centre of the circle represents white. Newton had shown in previous experiments that all the spectral colours combined give white light. Therefore if the little circles p, q, r, etc are proportional to their spectral brightnesses, they must yield white at the centre. The length of the line from the centre of the circle to the colour at Z then measures the tint or degree of saturation of the colour; if the resulting mixture should fall on the circumference itself it is a fully saturated spectral colour while if it falls on O it is white. Saturated colours then have no white mixed with them; desaturated colours such as pastels have a degree of white intermixed to give them a pale appearance. From the argument of finding colour mixtures by centres of gravity, all compound colours give points inside the circle. Points outside the circle have no physical interpretation in terms of spectral colours. Finally, Newton pointed out that colours directly opposite each other on the circle should be complementary; that is, colours whose combination should yield white at the centre of the circle. But no two of the little circles in Newton's illustration are directly opposite each other. And Newton himself said that 'I could never yet by mixing only two primary Colours produce a perfect white.' [5, p156] But that it might be produced from three primary colours he was not sure. In any case this difficulty was pointed out by Biot and became the subject of a mathematical investigation by Hermann Günther Grassmann in 1853 to be discussed in Chapter 6.

Compounding white from combined spectral colours was a unique idea which Newton spent much effort justifying. Reasonably enough he extended the idea to colours more commonly known and easier to work with, ordinary painter's powders. [5, p150] Here too he attempted to mix combinations of orange, purple, green and blue which 'became of such a gray as verged to no one of the colours more than another.' [5, p151] Newton approached an understanding of subtractive colour mixing when he explained that

> All coloured Powders do suppress and stop in them a very considerable Part of the Light by which they are illuminated. For they become colour'd by reflecting the Light of their own Colours more copiously, and that of all other Colours more sparingly, and yet they do not reflect

the Light of their own Colours so copiously as white Bodies do. [5, p150]

Therefore we could not expect a 'strong or full white'. Still, Newton in analogy with his spectral observations, had expected to produce white from his pigment mixtures and the statement above probably expresses Newton's surprise at not having done so. Within this explanation, however, lay an apparent solution to the difficulty; by simply increasing the illumination on the darkened colours they will indeed appear white. To demonstrate this, Newton spread a grey powder mixture on the floor of his room directly under strong sunlight. Now it appeared intensely white. [5, p152] In fact by suitable adjustment of the intensity of sunlight on the powder and on a piece of white paper, Newton made them 'both appear exactly alike in Whiteness.' [5, p153] Therefore, Newton satisfied himself that both spectral and pigmented colours should give white when mixed in proper proportions. And so was born a long-lasting confusion. Some commentators agreed with Newton's position while others argued (correctly) that mixtures of red, yellow and blue paints should ideally yield black. All of them, except one, argued like Newton for pigment mixtures from an analogy with spectral colours, and none of them, apparently, repeated Newton's experiments.

Unfortunately in viewing his powder mixture under strong sunlight Newton was apparently subject to a physiological effect not explicitly noted for almost 150 years afterwards. Any colour, whether monochromatic or mixed, will appear dazzlingly white if it is bright enough. Under strong sunlight Newton's grey would have had its reflected brightness raised sufficiently to stimulate this response in the eye. But had Newton viewed this reflected light through a prism, which he did not, it would not have yielded a spectrum comparable to that of direct sunlight. The confirming 'white' which Newton referred to was ultimately an artifact of the eye. Clearly Young was also a victim of this subjective effect when in emulating Newton he believed he produced a 'perfect whiteness' with powders viewed in strong sunlight.

As stated before, the circle procedure for mixing colours was given by Newton as an empirical rule only. It simply serves to organise the colours for producing combinations. It allows compound colours to be defined operationally in terms of angular positions along the circumference of the circle and linear distance from the centre. In effect, a colour is specified in terms of polar coordinates. As we shall see, Newton's rule exercised the main influence on most subsequent attempts at producing a quantitative theory of colour. It was the basis for work by Tobias Mayer and J H Lambert later in the eighteenth century and J D Forbes, Hermann Günther Grassmann and James Clerk Maxwell in the nineteenth century. Maxwell's method of colour equations (see Chapter 8) which still forms the basis for many colour measurements is a direct descendent of Newton's colour circle.

Newton's colour theory had an important and unlooked for influence upon the artist–printer Jakob C LeBlon. Born in 1667 in Germany, LeBlon was an early enthusiast of Newton and attempted to put into commercial practice what Newton (and possibly Waller) spoke about conceptually. Emigrating to London in 1720, LeBlon set up a stock company to print colour reproductions of famous paintings. These he planned to make by overlaying separate printings in red, yellow and blue ink. Describing the process he wrote:

> Painting can represent all visible objects, with three colours, yellow, red and blue; all other colours can be composed of these three which I call Primitive . . . Mixture of these three original colours makes black and all other colours whatsoever . . . I am speaking of material colours, or those used by Painters; for a mixture of all the primitive impalpable Colours, that cannot be felt, will not produce black, but the very contrary, White; as the great Sir Isaac Newton had demonstrated. [6]

Most remarkable here is a clear distinction between mixing pigment colours and mixing colours of light. In this he exceeded even Newton and all his successors for the next 120 years; for as we have seen, part of Brewster's difficulties lay in confusing the two processes.

LeBlon's venture was a technical success although he soon found that prints made by overlaying the 'primitive' colours only did not print satisfactory blacks. Hence he resorted to a fourth plate for the blacks, a modern practice even yet. Unfortunately the company went bankrupt due to inept management, but the few surviving LeBlon prints testify to his abilities and the basic truth of his theory.

In contrast with LeBlon, Moses Harris, a British entymologist, attempted a Newtonian circular organisation of colours in 1766. Using red, yellow and blue paints, Harris arranged his hues in two separate circles, entitled 'Prismatic Colours', and 'Compound Colours' (see Plate 4). It was as Harris said an attempt

> to discover all the variety of colours which can be formed from Red, Blue, and Yellow; which three grand or principal colours the author presumes do contain all the colours and teints in nature when mixed or blended with each other in the various proportions of their powers. [7, p3]

Harris's 'prismatic circle' in particular starts with the three primary colours arranged on the circumference of a circle together with fifteen intermediate colours compounded from them including purple. Each colour gradually faded into the next one along the circumference so there could not 'be a colour in the whole circle composed of more than two of the others'. [7, p5] The eighteen separate colours, were darkened in twenty steps inwards towards black at the centre of the circle yielding 360 colour variations. That the three primaries together should yield black he emphasised by illustrating at the centre of the circle three triangular pieces of stained glass in red, yellow and blue whose points overlapped to give black. Unfortunately, Harris was rather confused. For although he borrowed the circle idea from Newton, Newton had required white at the centre of his circle based upon his spectral experiments; contrary to Newton also, a distance from the centre of Harris's circle measured the brightness of the colour, not saturation. From Harris's colour circle of 'Prismatic Colours', it appears that he thought mixtures of paints behaved the same as mixtures of light. He then doubly confused the issue by requiring the pigment primaries to make black rather than white. In that he was, of course, correct but he was then incorrect to borrow Newton's circle, for that instrument is basically an additive device.

In any case Harris then went on to his second colour circle, that of 'Compound Colours'. Here, three 'intermediate' colours taken from the first circle, orange, green and purple, were arranged as second-order primaries on the circumference. Since each intermediate primary colour was originally a compound of two primaries, the colours now being compounded were the result of all three original primaries. Then as before, each hue was darkened in twenty steps inwards towards black at the centre yielding 300 new colour variations. (The twenty variations of each of the three mediates had already been counted in the primary circle.) Most extraordinarily Harris actually hand painted each separate hue in its appropriate place on the two plates, which resulted in very few copies of the book being printed. The work is probably more a monument to Harris's infinite patience than to his technical success. Yet despite its limited circulation it was known to other artists although not to members of the scientific community. Indeed until the later nineteenth century there were really two separate traditions of colour organisation, one by men associated with the arts and the other by men concentrating on scientific questions. Although there were necessarily similarities between the two, mostly they merged only when men whose professional interest lay in science also took an interest in colour. We shall now look into several examples of this tradition.

The eminent German mathematician Tobias Mayer produced a colour classification in 1758 that to some extent is still in use. Mayer was probably influenced toward this particular work by a combination of three motives. [8, p99] He was an extremely talented artist whose early precocity had played an important role in his youth. Secondly, he was a noted cartographer and saw a need for greater colour consistency on maps. Finally, the scientific side of this many faceted man may well have influenced him to attempt to reduce art to mathematical terms, especially in view of his personal philosophy to strengthen practice by theory. He even considered that young men trained in his system would be useful to book publishers in preparing colour illustrations. [8, p112]

As his primary colours Mayer chose red, yellow and blue, a choice that we have seen was already accepted in Boyle's time over 90 years earlier. But here Mayer differed from his predecessors for he aimed at giving the specific ratios of the primary colours required to mix all others; on this his predecessors had been satisfied with only qualitative ideas. Even LeBlon had not concerned himself with this detail although it seems that exactly this problem must have contributed to his financial difficulties.

With the objective of providing quantitative ratios for the compound colours it was necessary for Mayer to standardise the primary colours. This new consideration raised a still more fundamental question. In viewing transitions from one colour into another, say from red to blue, each separate step must be distinguishable but not so large as to appear a sudden change. In other words, transitions between colours were to be by the smallest sensible stages. Therefore, in the transitions between red and blue, the question was: How many small changes can be made, one at a time, so that each is gradual yet distinguishable? Mayer felt this number was 12, so this number became the standard amount for the primaries. [9] This means that in colours mixed from the primaries, the sum of their separate parts must equal 12. For example, using r, y and b for red, yellow and blue respectively, one could have combinations of r^6y^6 or $y^{11}b^1$ as (arbitrary) examples of binary combinations and $r^9y^2b^1$ or $r^6y^3b^3$ as examples of tertiary combinations. Mayer allowed variations of one unit of colour, but not less than one, in his scale of 12 and compiled a list of all possible combinations of either two or three primary colours. This totalled 91 (including the primaries) which it turns out is a triangular number of side 13, a fact recognised by Mayer. Therefore he said that each primary colour could represent the vertex of a triangle from which all the colour combinations could then be derived. Imagining weights at each vertex of the triangle, compound colours were located as in Newton's colour circle by finding the centres of gravity of combinations of weights (see Figure 4.2). A side of the triangle between any two primary colours

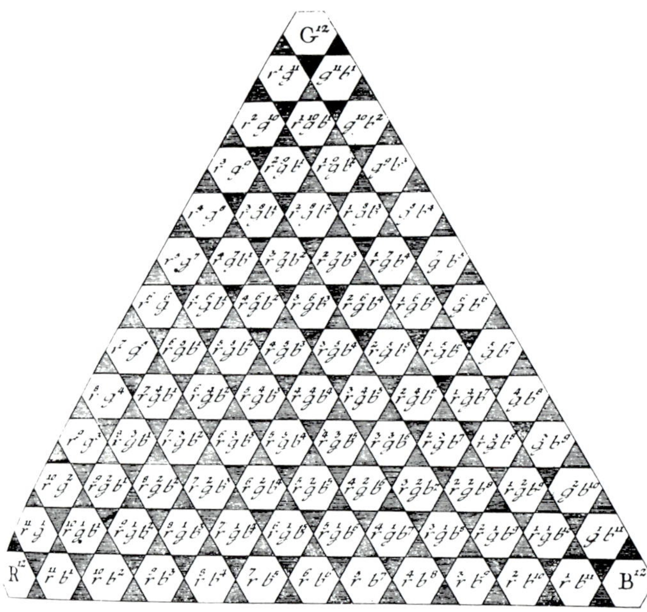

Figure 4.2 Triangular classification of colours by Tobias Mayer using red, yellow and blue primary colours. Colour graduations take place in steps of one part on a scale of twelve parts. (Reproduced from *Tobias Mayer, Opera Inedita: The First Translation of the Lichtenburg Edition* by Eric Forbes, by permission of Macmillan (London and Basingstoke).)

represents gradations between that pair of colours and in going around the perimeter of the triangle we must pass through all the spectral colours (by implication) and also purple. Moreover, as in Newton's circle, combinations of three primaries must fall within the triangle, colours outside the triangle having no meaning. Clearly, Mayer's triangular colour classification is a reduced form of Newton's circle although it does not yield white at the centre of the triangle. There the colours are rather dull, verging towards grey as expected from Mayer's choice of primaries. In any case, this is the origin of the famous triangle of colours used later by Young and Maxwell and still found in elementary science textbooks. An actual implementation of the triangle using six colour divisions is shown in Plate 5. [10]

The 91 different colours which this scheme allows seems rather small compared to nature's multiple hues. The reason is that the system has not allowed for shades or tints of the various hues. We have compounded only varieties of the (ideally) pure saturated primaries. But we can have gradations of each compound colour or primary towards white by mixing in a white pigment; and we can have gradations towards black of each hue by mixing a black pigment with the primaries.

Mayer recognised this requirement and allowed the primaries to move towards white to produce desaturated tints. On the scale of 12 he diluted each primary colour towards white by one step so there was one part white plus eleven parts colour. This resulted in fewer separate colours for as the primaries were diluted, the colours compounded from them became less distinguishable. In effect, the colour scale became smaller with larger steps between transitions. The diluted primaries then formed a new triangle containing 78 separate colour elements. Then one could repeat the process: dilute each primary with a second part of white so each new 'primary' consists of two parts white and ten parts colour. A new triangle with these now forms 66 separate colour elements. Continuing in this manner, there will be altogether $91 + 78 + 66 + 55 + \ldots + 1 = 455$ colour elements, a number which Mayer recognised as a pyramidal number of side 12. The base of the pyramid is the original triangle and its apex is white. A similar argument applied to darkened shades yielded another pyramid with a black apex. There are therefore altogether 819 separate colours presumably sufficient to represent a great variety of subjects. In the sense that artists refer to colour differences in terms of hue, tint and shade, Mayer showed that three dimensions of space are necessary to account for separate colours. This elegant theory unfortunately never came to fruition for Mayer encountered numerous practical problems in implementing it, such as the physical and chemical properties of the pigments, the solvent materials to be used, etc. All these detracted from his purpose of reducing art to a simple, practical form. As LeBlon had discovered, it turned out to be less simple in practice.

The theory given here was announced in a lecture in 1758 but only published in 1775, thirteen years after Mayer's death. In the meantime, Mayer's distinguished countryman, Johannes Lambert, independently discovered exactly the same system which he published in 1772. [11] [17, p171]

Suffice it to say that Lambert visualised placing each diluted triangle on top of the previous one until a pyramid with white at the top was obtained. Similarly there would be a pyramid with black at the apex. Placing both tetrahedrons together at their bases formed a double pyramid, white at one apex and black at the other with the original triangle of colours half-way between. Similar attempts at colour classification were made in the nineteenth century (and the twentieth) together with attempts at fixing a standard colour nomenclature. We shall briefly look into a few of these in the first half of the nineteenth century.

In 1810 the German romantic painter Philip Otto Runge devised the colour sphere shown in Plate 6. [12] Runge too chose red, yellow and blue primaries, arranging them with their gradations into each other around the equator of a sphere. His equator then was analogous to Newton's colour circle. At opposite poles Runge placed black and white with all greys compounded from them on a diameter between them. Then mixed colours were located inside the sphere or on its surface, with gradations towards white at one pole and gradations towards black at the opposite pole. Unlike Mayer and Lambert, however, Runge attached no step-wise variations to the colours, nor proposed any standardisation of the primaries. Colour variations were continuous so in effect the sphere contained an infinite number of colour variations. Therefore his colour organisation is only a qualitative description but given this it clearly allows for a complete colour classification.

Colour classification passed from theoretical exercises to an important practical application with the work of M E Chevreul in 1839. This celebrated French chemist was then director of dyeing at the *Manufactures Royales des Gobelins*, the national tapestry workshop. Succeeding to the position in 1824, he was first assigned the seemingly chemical problem of improving dye intensity and fastness. These problems Chevreul did solve, as expected, but complaints still arose concerning the apparent low intensity of darkened blues and violets. Convinced that the colours were unfaded and as bright as intended, Chevreul began studying colour impressions in terms of contrast. Here he made a capital discovery, first announced in 1839, that the apparent brightness of a given colour depends more importantly on the colours surrounding it than on the intensity of the dye itself. This discovery he called the law of simultaneous contrast stating that:

> where the eye sees at the same time two contiguous colours, they will appear as dissimilar as possible, both in their optical composition and in the height of their tone. We have then, at the same time, simultaneous contrast of colour properly so called and contrast of tone. [13, p11]

It was an extension of this work that led Chevreul to colour classification with the objective of describing the possible changes of colour. He arranged the colours along the circumference of a circle with his three primary colours, red, yellow and blue, spaced 120° apart. Referring to Figure 4.3, each radius was divided into 20 parts representing gradations of that colour from white at the centre to the 'pure tone' of that colour somewhere along the radius

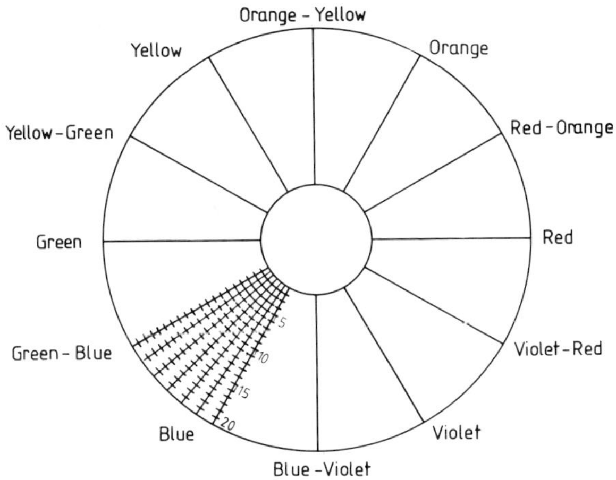

Figure 4.3 Chevreul's circular colour classification showing only one sector completely.

although the saturated colour was not necessarily on the circumference. This is shown for simplicity only for blue. Chevreul arbitrarily chose 15 as the value for the pure (saturated) hue of the red scale stating that because of differences in reflectances, the number chosen for yellow would be lower while that for blue would be higher. Then each colour was diluted by white in steps towards the centre of the circle with increasing whiteness representing increasing brightness. The centre of the circle therefore represents white which can be produced by one colour only of sufficient intensity and does not represent the addition of colours to make white in Newton's sense. Opposed to this, moving towards the circumference of the circle each colour was darkened but by how much is vague. [13, pp58–9] For the gradual transitions of one primary colour into another (i.e. combinations of two primaries) he divided the angles between the primaries providing for all the spectral colours and their intermediate hues. This is shown in Figure 4.3 where for simplicity these are shown for one sector only. [14] As can be seen, there are within the circle (as completely drawn by Chevreul) $72 \times 20 = 1440$ colours representing the primaries, their intermediate combinations and their tints towards white.

For gradations into black Chevreul employed a quadrant of radius equal to the circle and placed so that it could turn around an axis perpendicular to the centre of the circle. The quadrant is shown in Figure 4.4 and the hemisphere resulting from it in Figure 4.5. The quadrant is divided into 10 radii giving various degrees of black mixed with the colour lying below it in the circle. The smallest degree of black is represented by radius no 1 labelled 1N, 9C for one part noire (black) and nine parts colour. Radius no 9, labelled 9N, 1C represents nine parts black and one part colour while radius no 10 represents the maximum amount of black. Radius no 10 is in fact the axis about which the quadrant revolves. It is pure black at the top and of necessity white at the

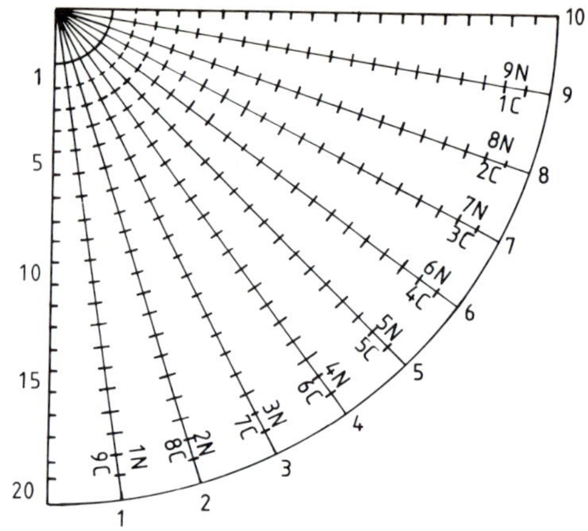

Figure 4.4 Chevreul's quadrant for shades of black to darken each of the colours on the circle. The quadrant rotates about the axis labelled 10 which is perpendicular to the circle.

bottom, and therefore represents 20 degrees of grey considered as white and black mixed.

To mix each colour with black, the radii of the quadrant are divided into 20 parts so that as it sweeps around, any colour on the circle has a point of the quadrant over it. There are therefore 1440 × 9 = 12,960 darker variations of the original colours along the quadrant radii and 20 grey combinations along the 10th radius. Altogether then Chevreul's hemisphere contains 14,420 colour variations, more than enough certainly for any artist but grossly impracticable to construct.

In requiring three dimensions of space, Chevreul's classification was similar to those of Mayer, Lambert and Runge. However, Chevreul arrived at his scheme independently, seeing himself as expanding Newton's colour circle rather than improving their designs. Nevertheless his goal was similar to theirs: to standardise the definition of colour. It seems an elaborate and complete system of colour classification, but there are even more problems

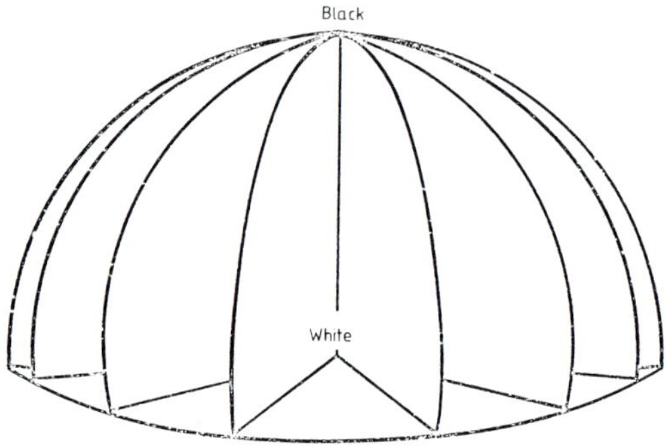

Figure 4.5 Chevreul's colour hemisphere.

with this method than with theirs. Those systems were intended as mechanisms for compounding many colours from three primaries. Chevreul, however, ignored that idea, for he did not show how intermediate colours on the circle were compounded from his primary colours. He had simply allowed variations in luminosity of each colour in the circle. Therefore by implication, each colour intermediate must be found as a separate pigment, not as a compound from the primary colours themselves. On the other hand if other colours are to be compounded from the primaries then the centre of the circle must be black instead of white and Chevreul clearly recognised that mixture of red, yellow and blue yields black ideally. [13, p57] But his centre is white due to dazzling luminosity rather than representing an addition of primary colours in Newton's sense. So there is a contradiction in terms. If one mixes colours from the primaries, the centre of the circle must be black. If the centre remains white, one cannot compound intermediates from the primaries and they ultimately serve little purpose. The latter is Chevreul's own understanding. The circle in the end does not represent mixed primary colours but only varying brightnesses of every separate colour each of which must be represented by a separate pigment. Moreover, similar (black-containing) mixtures can occur twice in the hemisphere so Chevreul's organisation ultimately violates the cardinal rule of colour organisation that each hue corresponds to one and only one point in space. Not surprisingly considering its complexities the hemisphere was never constructed at the *Manufactures Royales des Gobelin*, the very place of its birth.

If the colour hemisphere was not very useful, Chevreul's law of simultaneous contrast was well received and this together with his other colour work made him one of the most influential colourists of his time. Not only did he solve the chemical problems of dyes, his original problem, but he vastly improved the tapestry designs by colour selections based on his law of simultaneous contrast. Indeed that idea much influenced the work of later impressionist painters who insisted that the effect of each colour must be judged in terms of the adjacent colours.

COLOUR NOMENCLATURE

The mere classification of colours according to schemes such as these were thought insufficient for general scientific purposes; general agreement upon colour nomenclature was also needed. Not that language was deficient in colour vocabulary; rather there was simply confusion over the meaning of many colour words, suggesting that many definitions were not well fixed. This lack of agreement naturally enough led to seeming misapplications of colour terms. Hence there was a perceived need for some standardised colour vocabulary, especially in the descriptive sciences where complete specifications included colour. For example, the German mineralogist and geologist Abraham Werner brought mineral classification to a new understanding in this period in 1774, but he found the existing colour vocabulary insufficient for his mineralogical requirements. Therefore, he established his own nomenclature of 52 colours as an integral part of his mineralogical classification.

For more general purposes 52 colours are certainly insufficient when compared with the 819 hues produced theoretically by Mayer and Lambert to say nothing of the potentialities in Chevreul's system. Recognising this limitation, Patrick Syme, a noted artist and designated painter to the Wernerian Society in Edinburgh, attempted in 1821 to extend Werner's nomenclature. [15] Noting that names of colours were still often misapplied, he aimed at removing the general confusion and at producing a system useful to the descriptive sciences. His book was based on Werner's system but extended the number of terms to 110, still rather too few. For this reason his work achieved little success. In the end attempts at defining colours ran into the very difficulties precipitating them. Each artist depends more on his own subjective feelings than on someone else's terms so colour vocabulary continued to reflect the general disagreement in terms. Thus lack of success on Symes' part did not indicate lack of appreciation of his work.

Of all attempts at naming colours, that which had the greatest success was undoubtedly produced by D R Hay in 1845 entitled *A Nomenclature of Colours*. This Scottish decorator and artist was well known in the thriving intellectual community of Edinburgh for both his outstanding talent and his numerous publications on the relations between art and science. His organisation of colours together with appropriate names fell on fertile ground in the persons of his eminent countrymen, James David Forbes and James Clerk Maxwell: of these, much more later.

Hay's objective was the high ideal of improving the education of the general public in 'aesthetical science'. [16] But the basis for this he thought was first education in knowledge of colours. And so he aimed at 'a definition of the almost infinitely various colours, hues, tints, and shades' [16] arising out of the simple elements of colour. These elements were red, yellow and blue and three secondary colours, orange, green and purple. Finally black and white were not to be considered colours as such but as modifiers of colour in the sense of changing them into tints (addition of white) or shades (addition of black).

The major portion of Hay's work consists of a series of colour plates such as those in Plate 7. Generally two plates are considered together. Plate 7 (*right*) consists of tints of red contrasted with shades of green all mixed according to specific ratios. However, he does not state the actual pigment material used. Plate 7 (*left*) exhibits tints of green contrasted with shades of red again according to specific ratios. In each plate a colour in the upper half diluted with white (a tint) is contrasted with the colour directly underneath it in the lower half darkened with black (a shade). Besides exhibiting plates of primary and secondary colours alone, Hay also arranged many more such plates combining primary colours, or secondary colours, or both primary and secondary colours, etc, always in terms of contrasts between their various shades and tints. In all he achieved some 240 different varieties and only the expense of the colour lithography (remarkable for the time) prevented him from completing his system with 144 additional colours. It was in short a noteworthy work, no less so for the considerable attention it drew to itself as for its influence on Forbes and Maxwell, as noted.

J D FORBES

The eminent Scottish physicist, James David Forbes, produced in 1849 probably the most important work in colour theory prior to Helmholtz and Maxwell in the 1850s. [17] It was primarily a comprehensive review of the status of colour understanding with a detailed summary of the work of Mayer and Lambert. But Forbes also made an original contribution with an analysis of an old problem.

The confusion lay in deciding whether the mixed colour primaries should produce either white or black. This difficulty in turn was traceable to Newton who, of course, opted for white. In succeeding years LeBlon, Harris, Mayer and Lambert had thought of their primaries making black. But they also considered their colour mixtures in analogy with spectral colours. Only LeBlon had made a clear distinction and his short remark was unknown to any of these men. The perplexity lay in the fact that the mixed primaries produced neither white, as Newton required, nor black as expected by others. Mixing pigment colours in analogy with spectral experiments gave at best a neutral grey which could be taken to confirm either opinion. The failure to compound white from the three primary colours (as viewed by the physicists) remained unsolved. It was in this context of repeated failure by Newton, Young, Hay and others that Forbes asked: Why is the best possible result of mixing pigments to make white, always a neutral grey? And, why can we not produce white from the spinning disc method although we can always achieve grey? Assuming that mixed primaries should give white, Forbes first attempted to distinguish the two processes of colour mixing underlying the confusion. Then with a numerical analysis he corroborated, albeit hypothetically, Newton's explanation of grey instead of white from pigment mixtures.

The first point made by Forbes was that mixing coloured lights projected on a screen is an additive process where the resultant brightness is the sum of the individual brightnesses. But mixing paints is a subtractive process due to absorption and scattering processes which produce 'an inevitable quantity of darkness or absorbent faculty in the constitution of every artificial colour, whatever be its predominant hue.' [17, p165] For example, suppose we have a paper coated with blue pigment and another coated with yellow. Suppose they reflect only their own colours but at the same time absorb, say, half of the incident light of that colour. Then a compound colour of these two pigments spread on a surface must reflect, at best, only half of the total number of rays of those two colours. Therefore such a procedure cannot make a surface more reflective than before, but less. As Forbes stated:

> the process of mixture cannot in any case be expected to *improve* the power of reflecting the pure colours residing in the constituent pigments. It is much more likely to deteriorate it, which will tend to give a tone of always less absolute brightness to more complex colours. [17, p166]

This seems at first to be an elementary and even trivial point. But the fact that Forbes, a well known and respected physicist, felt called upon to make such a statement is indicative of the general confusion then concerning the physical mechanism of colour mixing. In fact, the use of the word 'sum' in conjunction with mixing coloured lights implies that Forbes was groping at least toward a quantitative as well as physical conception of that process. Forbes did not, however, ever use the word 'subtractive' in referring to pigments but the inference about the mathematical operation is clear. Yet in the end, as we shall see, he actually missed the distinction between the two processes.

The discourse above was preliminary to getting to an actual example based upon real measurements. Forbes' experimental method consisted of that tried and true procedure for combining colours, the revolving disc, but this time with a clever modification. Whereas others had painted the colours on the disc he used overlapping coloured papers (provided by Hay) secured by a screw at the centre of the disc. Therefore the colour areas were easily adjustable. Pursuing the question of colour combinations making grey he attempted to obtain equations between certain mixed colours and grey. [18] Here he was successful as expected with the red, yellow and blue primaries but to his surprise he also obtained an equivalence between yellow and blue and grey. [19] Since yellow and blue should have yielded green he considered this an anomolous result. However, he ignored this unexpected result being more interested in the reduced light intensity yielding grey than the resultant colour itself. Hence he missed the chance to make an important discovery. He concentrated on the greyish combinations of red, yellow and blue, always assuming these were the correct primary colours.

Suppose now we wish to have a spinning disc with sectors of red, yellow and blue to make white. The question is: What proportions of the disc should be assigned to each colour? The correct proportions could be found from experiment, i.e. trial and error, which seems to have been Young's method. But they can also be found from Newton's colour circle. There the little circles q, p, r, etc represent the centres of gravity of the colour sectors red, yellow and blue respectively. Using these as reference points, Forbes determined the angle between the centre of gravity of red and yellow as 91° 54′ 30″, that between yellow and blue as 115° 26′, and that between blue and red as 152° 39′ 30″. (How he achieved such accurate measurements from Newton's small circle is hard to see.) To place the centre of gravity of the whole circle at its centre, each colour is made proportional to the sine of the angle opposite it. Point p, for example, is approximately opposite the 115° 26′ angle. From the above proportions and measurements we have:

$$\begin{aligned} &\text{red} && \propto \sin 115° 26' = 0.9031 \\ &\text{blue} && \propto \sin\ 91° 54' = 0.9785 \text{ (Forbes' value, but correctly 0.9995)} \\ &\text{yellow} && \propto \sin 152° 39' = 0.4593. \end{aligned}$$

If X is the amount of yellow, then from these proportions $2.13X$ is the amount of blue, and $1.96X$ is the amount of red. Together they must add up to the 100 divisions of the wheel so $5.09X = 100$ and we have the proportions for the wheel as

$$\begin{aligned} &\text{red} && = 38.5 \\ &\text{blue} && = 41.8 \\ &\text{yellow} && = 19.6. \end{aligned}$$

Now let the circumference of the disc be divided into 100 parts numbered from 0 to 1.00. These numbers will be measured there as 0.385, 0.418, and 0.196 respectively.

This method, however, gives proportions for mixing 'spectral' colours assuming perfect reflection. But surfaces do not reflect 100% of the incident light even of their own colour, a fact well known to Newton. Lambert had published in 1772 estimates of the proportions colours do reflect and these were used by Forbes. [11] [17, p166] The whitest surface reflects 4/10 of white light incident upon it. Yellow reflects 4/10 of its own colour, the brightest red reflects 1/3 of its colour, and blue reflects 1/7 of its colour.

In order to obtain reflected light from the disc of the original brightness we must increase their proportions inversely as their reflecting abilities. Therefore, we must make the spaces on the wheel

$$\begin{aligned} &\text{yellow} && 10/4(0.196) = 0.490 \\ &\text{blue} && 7(0.418) = 2.926 \\ &\text{red} && 3(0.385) = 1.155 \\ \hline &\text{white} && \text{sum} = 4.571 = n. \end{aligned}$$

In other words to match the original brightness we would have to make the wheel 4.57 times larger than originally planned. Conversely, if we keep the proportions of the wheel as determined only by Newton's colour circle, the arithmetic shows that the reflected light will be attenuated 4.57 times becoming $1/4.57 = 0.22$ of its original brightness. But the same white light falling on white paper will have $4/10 = 0.40$ of its original value reflected. Comparing these two we see that the 'white' light reflected from the coloured wheel as compared with white light reflected from the white surface is about half. Forbes was satisfied saying 'this abundantly explains why the result is a grey colour, not a bright white.' [17, pp167–8]

Neither Newton nor Young might have been surprised at this conclusion but it remained for Forbes nevertheless to quantify what they only spoke about. We may now understand why Young's spinning disc using red, green and blue only yielded grey. Unfortunately, Forbes continued to use yellow in this plausibility argument and in doing so he ironically brought the story 'full circle'. His concept of a subtractive process in colour mixing was restricted to the reduced light intensity by reflection only. He did not consider subtractive effects (i.e. absorption) on the resulting hue itself. So if we can understand why Young's disc yielded only grey we can also understand why Forbes' disc yielded only grey. As we have seen mixtures of red, yellow and blue argued for neither white nor black; therefore, if Forbes had assumed these colours produced black rather than white he could have made a similar argument for them still producing grey. Forbes' basically correct analysis notwithstanding, nothing much has changed.

It is interesting to view against the background of colour classification the choice of primary colours of Young, Wollaston and Brewster. Whereas the received opinion that red, yellow and blue are primaries goes back at least to Boyle in 1664 and was carried over to the choice of spectral primaries, Young and Wollaston at least made a case for a different choice. Indeed, Forbes remarked that Young's choice was a 'singular opinion' [17, p172] indicating that in 1849, 47 years after its announcement, Young's choice of green over yellow as a primary colour remained quite neglected if not forgotten. This was partly due to the uncertainty of Wollaston's observations. Young had inferred from that work that yellow does not exist independently in the spectrum but is composed of a mixture of red and green. But the independent existence of yellow in the spectrum had been readily shown by Brewster and John Herschel in their absorption experiments while subsequent work by Fraunhofer indicated the inadequacy of Wollaston's original observations. Therefore, the original reason for challenging the general opinion no longer existed.

A more powerful reason for rejecting Young's choice of green lies in the fact that it can be fitted into the red, yellow and blue scheme. One assumes, as Brewster did, that red + green = orange. Then Young's statement that red + green = yellow can be explained by assuming that green is a compound of yellow and blue. According to Forbes, 'the whole of the blue and a part of the yellow combine with the red to produce a perfect white, which then dilutes the outstanding portion of the yellow.' [17, p172] Thus Young's choice of red, green and violet spectral primaries was seen as a special case of the then more conformable choice of red, yellow and blue.

Returning to Forbes review of colour theory, he then analysed the composition of certain colours constructed from a triangle like Mayer's but using eight divisions rather than twelve. The figure eight was suggested by some rather questionable experiments made in 1820 and 1835 by the chemist George Field. [20]

Referring to Forbes' triangle in Figure 4.6, the primary colours at the vertices can be considered to form secondary colours at the midpoints of each side of the triangle. Connecting those points gives another, smaller triangle which can be used for compounding tertiary colours. These are the midpoints of the smaller triangle of secondaries. The process can be continued producing a great 'confusion of colours' [17, p174] with decreasingly smaller triangles. The centre, W, is the limit of the process and we can see that relatively soon we will reach gradations which will hardly affect the eye.

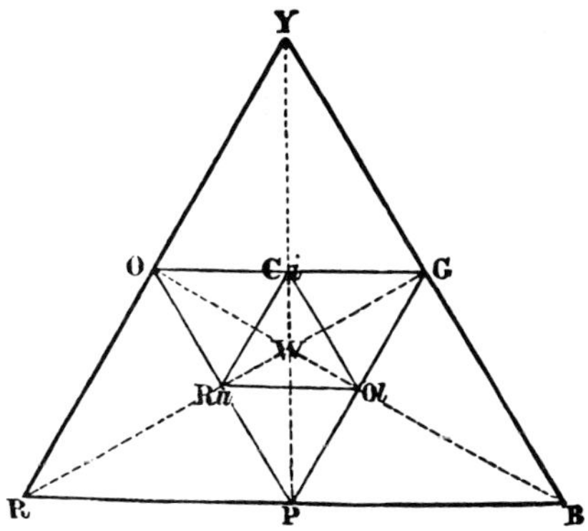

Figure 4.6 Colour triangle of J D Forbes (1849) showing successive additions of colours starting with red, yellow and blue primaries. Addition of two primaries yields a secondary colour lying between them. These points become the vertices of a new triangle of colours and the process is repeated yielding white as a limit at the centre of the RYB triangle.

Forbes illustrated his idea of combining secondary and tertiary colours as shown in Figure 4.7, still basing their composition on the arbitrary sum, eight. [17, p174] The colours on the left are the primaries, the next column are the secondaries compounded from them, the next column the tertiary colours, etc. This classification scheme owes a debt to Field's work but is vastly simpler.

The colour citrine seems to be an indefinite one. It is composed of four parts of yellow and two parts each of red and blue. But since grey (ideally white) is a mixture of red, yellow and blue, this is equivalent to mixing two parts yellow with six parts of neutral grey to make the total of eight parts. This shows that citrine is a 'yellow verging into grey, one-fourth of the mixture being yellow and three-fourths grey.' [17, p174] Consequently its place in the triangle must be on the line YW, dividing the line in the ratio of 3:1. It does this by construction.

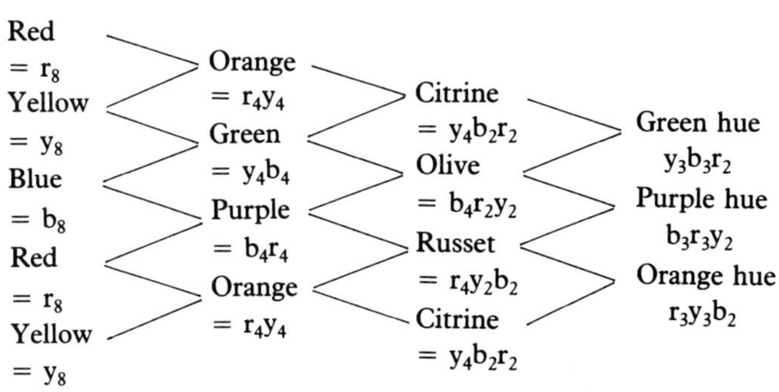

Figure 4.7 Successive composition of colours as derived by Forbes from the triangles of Figure 4.6.

Table 4.1. Colour classification of J D Forbes

Red	Greyish-red	Grey-red [Russet]	Red-grey	Reddish-grey	Grey
Orangish-red	*	*			
Red-orange	*	*	*		
Reddish-orange	*	*			
Orange	Greyish-orange	Grey-orange [Brown]	Orange-grey	Orangish-grey	Grey
Yellowish-orange	*	*			
Yellow-orange	*	*	*		
Orangish-yellow	*	*			
Yellow	Greyish-yellow	Grey-yellow [Citrine]	Yellow-grey	Yellowish-grey	Grey
Greenish-yellow	*	*			
Yellow-green	*	*	*		
Yellowish-green	*	*			
Green	Greyish-green	Grey-green [Drab]	Green-grey	Greenish-grey	Grey
Bluish-green	*	*			
Blue-green	*	*	*		
Greenish-blue	*	*			
Blue	Greyish-blue	Grey-blue [Olive]	Blue-grey	Bluish-grey	Grey
Purplish-blue	*	*			
Blue-purple	*	*	*		
Bluish-purple	*	*			
Purple	Greyish-purple	Grey-purple [Slate]	Purple-grey	Purplish-grey	Grey
Reddish-purple	*	*			
Red-purple	*	*	*		
Purplish-red	*	*			

Russet intermediate between red and grey.
Brown intermediate between orange and grey.
Citrine intermediate between yellow and grey.
Drab intermediate between green and grey.
Olive intermediate between blue and grey.
Slate intermediate between purple and grey.

Forbes performed a similar analysis of the still more indefinite colours listed in the fourth column. He showed that in the triangle composed of the points Ci (citrine), Ra (russet) and Ol (olive) the next set of colours in the fourth column also divided the altitudes of that smaller triangle in the ratio of 3:1. The only difference was that with these colours, 'the compounds were more neutral, because the secondary colours were themselves one stage on the way to neutrality.' [17, p175] From these examples and with the help of Hay in actually arranging the colours, Forbes came to the conclusion that:

> all combinations of three primary colours (as far as difference of quality is concerned) may be represented by transitions from the primary and secondary colours into grey; and thence, though it may appear at first sight paradoxical, though the quality of a primary or secondary colour (such as red or green) is not changed by diluting it with white, it is changed by mixing it with grey, or by first mixing it with white and diminishing the intensity of light in the mixture. [17, p175]

Clearly such a system of classification was useful in pointing out the composition of each colour. It did not, however, include the possibility of dilutions of these colours with black or white, the colours being viewed tacitly under a standard illumination. Following his classification scheme to its logical conclusion Forbes presented a table of colours using colours arranged for him by Hay from his *Nomenclature of Colours* [17, p175] (see Table 4.1). Forbes also attempted to match Hay's arrangements with painted porcelains and enamels from a unique collection in the Vatican. After long delay he managed to obtain 941 pieces from a collection of some 18 000. That result was disappointing but he was actually able to construct with them a representation of Mayer's colour triangle. [17, p178]

The value of Forbes' article was ultimately that it provided a complete résumé of the problems underlying classifying colours and defining primary colours. Although Forbes ultimately committed the mistake of his predecessors in confusing optical and pigmented colours, this article served an important function in bringing together in a convenient and well written form the existing knowledge of the time. The short historical résumé at the beginning of his article was closely followed by Helmholtz in his own paper on the subject in 1852. Maxwell read this paper and knew Forbes personally at the time of its writing. So Forbes' résumé together with his own contributions were an important stimulus to the next major advances by Helmholtz and Maxwell in the problem of understanding the nature of primary colours.

NOTES

1 Halbertsma, K J A 1949 *A History of the Theory of Colour* (Amsterdam: Sivets and Zeitlinger). Halbertsma attempts a history of the entire subject from the sixth century BC to 1949 in only 256 pages. The result is rather antiquarian

and at best is a capsular summary of events. He does, however, indicate the continuity of discussion on primary colours

2 Birren, Faber 1965 *A History of Colour in Painting* (New York: Rheinhold) p18

3 Waller, R 1685/6 A catalogue of simple and mixed colours *Phil. Trans. R. Soc.* **179**

4 Boyle, Robert 1664 *Experiments and considerations touching colours* (London) p220

5 Newton, Isaac 1730 *Opticks* 4th edn (London). Reprinted 1952 (New York: Dover)

6 LeBlon, Jakob C *Coloritto, or the Harmony of Colouring in Painting*. English and French Edition reprinted 1756 by Antoine Gautier de Montordage under the title *L'Art d'Imprimer les Tableaux* (Paris) pp28–30

7 Harris, Moses 1766 *The Natural System of Colours* (Licester-Fields) Facsimile reprint by Faber Birren 1963 (privately printed). The date of the original was estimated by Birren

8 Forbes, Eric G 1970 Tobias Mayer's theory of colour mixing and its application to artistic reproductions *Ann. Sci.* **26**

9 Mayer, Tobias *Von den Farben*. A German excerpt published in [8, p105]

10 Mayer, Tobias 1775 *De affinitate Colorum Commentatio* published posthumously in Georg Cristoph Lichtenburg *Opera Inedita Tobiae Mayeri* vol I (Göttingen)

11 Lambert, Johannes 1772 *Beschreibung einer Farben Pyramide* (Berlin) as described by J D Forbes [17]. The author has been unable to locate a copy of Lambert's work

12 Runge, Philip Otto 1810 *Farben-Kugel, oder Construction des Verhaltnisses aller Mischung der Farben zu einander, und ihrer vollständiger affinitat* (Hamburg)

13 Chevreul, M E 1860 *The Principles of Harmony and Contrast of Colours* 3rd edn (London). Translated from the original 1839 French edition by Charles Martel

14 The drawing is based on that given by Chevreul as a fold-out at the end of the *Atlas* appended to the 1839 original edition

15 Syme, Patrick 1821 *Werner's Nomenclature of Colours* (Edinburgh)

16 Hay, D R 1845 *A Nomenclature of Colours, Hues, Tints and Shades* (Edinburgh and London) p5

17 Forbes, James David 1848 Hints towards a classification of colours *Phil. Mag.* **34**

18 Letter of 4 May 1855 to James Clerk Maxwell published in Campbell, Lewis and Garnett, William 1882 *The Life of James Clerk Maxwell* (London: Macmillan) p214

19 Forbes, James David 1855 Observations on Mr Maxwell's Paper *Edinb. New Phil. J.* April. The observations were first communicated to Maxwell as a letter. He had actually hoped to pursue this observation but illness and other work prevented it (see [18])

20 Field, George 1835 *Chromatography; or a Treatise on Colours and Pigments, of their Powers in Painting, etc* (London). Many of the experiments described in this work, however, were performed as early as 1820. Much of its contents had appeared in the periodical *The Pamphleteer* **17** in 1820 under the title: Aesthetics, or the analogy of the sensible sciences indicated

5
Hermann von Helmholtz: New Insights into Primary Colours

So far in this history the underlying problem has been that of determining the composition of mixed colours. Hardly any new knowledge beyond Newton's existed on mixing spectral colours a few at a time and on this even Newton had enjoyed only limited success. Considering the wealth of publications on colour, colour classification and colour nomenclature, it is remarkable that by the mid-nineteenth century there had still been no concerted series of experiments on colour mixtures with lights.

To be sure the Belgian physicist J A F Plateau had found in 1829 that yellow and blue on a spinning disc did not make green, [1] but that work had been concerned with the duration of sight rather than colour sensation and the result went unnoticed. Forbes' discovery of this fact, of course, went unpublished. In Germany, Paul Volkmann had performed experiments in 1838 with a view toward making colour mixtures of light but his methods produced confusing results at best. With limited experimental guidance physicists (not to mention artists) who wrote on colour tacitly assumed that experience gained from art simply carried over to mixtures of light. Hence they always assumed that mixed colours of light behaved the same as mixed pigments. This we have seen from Newton in 1704 to Forbes in 1848. On this point Helmholtz said,

> this theory, however, has been based from the beginning upon a single mode of experiment; namely, that in which colouring substances are mixed together, the results being assumed to be the same as would follow from the union of the coloured lights themselves. [2, p524]

It was ultimately the extreme development of that assumption which allowed Brewster to conceive his triple spectrum theory that in turn became an obstacle to further progress. That theory was refuted in 1852 by Helmholtz who simultaneously recognised the tacit assumption and the need to investigate it experimentally. One month after the paper against Brewster was published, Helmholtz published a pivotal work on exactly this subject. There he demonstrated the correct means of thinking about colour mixtures with a series of experiments on mixed spectral colours such as never had been performed. We shall look into these in detail for they marked a turning point in experimental technique as well as in conception of colour.

Unlike his predecessors, Helmholtz experimented only with additive mixtures of light. Having grasped the error of his predecessors he set out to determine all possible combinations that could be made from mixtures of the spectral colours themselves. Helmholtz cut two narrow slits in a black screen each inclined to the horizontal at 45° and joined together to form a V. [2, p524] Thus they formed a right angle in the screen pointing downward. At a distance of 12 ft from the screen he placed his prism which in turn was in front of and close to a viewing telescope. The prism was placed in a position of minimum deviation, which position gives the sharpest spectral image and largest angular dispersion.

Now if one views a vertical slit through a vertically oriented prism, the result is a rectangular spectrum where each coloured band is vertical and Fraunhofer's lines are vertical (see Figure 5.1(*a*)). Now imagine that the

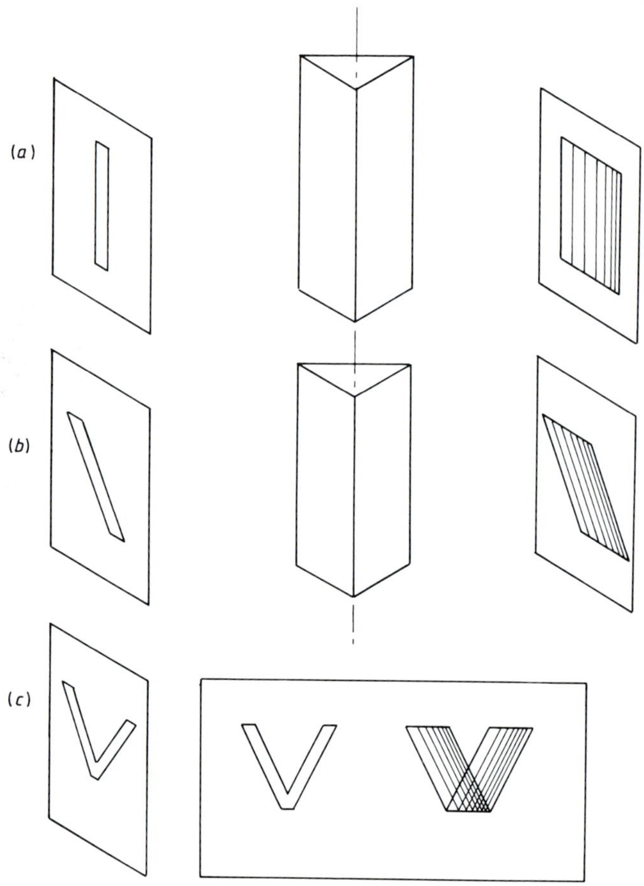

Figure 5.1 Slits, prism and resulting spectra based on descriptions by Helmholtz.

vertical slit rotates slightly to become an oblique slit. The spectral image is then 'an oblique-angled parallelogram, with two opposite sides horizontal and two others parallel to the slit' [2, p522] (see Figure 5.1(*b*)). The result is an oblique spectrum where the colour bands are always parallel to the slit. If then there are the two oblique slits forming a V, the spectrum formed from one leg will partly overlap the spectrum formed by the other leg. Since the angle between the slits is 90° in this case the two spectra intersect at right angles. The result as Helmholtz said is that 'every coloured band of the one intersects in the common field of the spectra each band of the other, and thus we at once obtain the total combinations capable of being formed out of every two simple colours' [2, p523] (see Figure 5.1(*c*)). [3]

Helmholtz used a highly dispersing flint glass prism in conjunction with a narrow slit illuminated by direct sunlight. He could see numerous Fraunhofer lines which he needed as reference marks, for it was crucial to the success of this technique that the colours in each separate limb be well separated. The Fraunhofer lines, as reference marks, assured him that no colour overlapping took place within each separate spectrum. The presence of the Fraunhofer lines had also another important purpose. Since the experiment was aimed at producing many binary colour combinations, it was necessary to determine exactly which two colours produced a given mixed hue. This was especially important considering that more than one set of colours can produce the same colour sensation in the eye.

In order to know which colours produced each mixture, Helmholtz mounted right-angled cross hairs in his viewing telescope. The cross hairs could then be aligned with the Fraunhofer lines in each separate spectrum which, of necessity, intersected at right angles. The cross hairs aligned on the Fraunhofer lines marked the two spectral colours producing a given compound colour. For illumination during actual observation of the colours, direct sunlight was not used as it could not evenly illuminate the full length of the slit. Therefore daylight or sunlight reflected off a white surface was employed.

While observing the mixed colours it was also desirable to alter the relative brightness of the individual colours producing the mixture. Then there was control over three variables in any compound colour; the two simple spectral colours of which it was composed and the brightness of one with respect to the other. This was accomplished by revolving the prism from a vertical position to one inclined to the horizontal. It was remarked above that given a vertical prism in front of a vertical slit, the spectral image is vertical, but upon rotating the slit, keeping the prism stationary, the spectral image becomes a parallelogram. Clearly then rotating the prism is equivalent to rotating the slit. By rotating the prism the parallelogram images maintain their sides of constant length but change their angles and this changes the area of the parallelogram. Since the area covered by the parallelogram is maximum when its angles are right angles, rotating the prism causes the parallelogram angle to become more oblique, occupying less space. But since the same intensity of light is still incident on the slit it then illuminates a smaller surface making the spectral image brighter. Turning the prism then changes the angle of each slit with respect to the prism, and one spectrum becomes brighter and the other dimmer.

This procedure, however, has difficulties. Referring to Figure 5.1(*b*) one can see that as the spectrum parallelogram becomes more oblique, it degenerates into a straight line causing the bands of colour to be crowded very close together. This will spoil the spectrum's purity and thus defeat the purpose of the experiment. Therefore, this procedure is rather limited and another method of producing intensity changes is also needed. This Helmholtz accomplished 'with great facility by placing pieces of paper, oiled or not oiled, of greater or lesser thickness, behind one of the slits.' [2, p524] Then he had a spectrum of diminished intensity from one slit together with the bright spectrum from the other.

There was, however, in all this still the danger of being deceived in one's observations through retinal fatigue due to bright light or through contrast effects as Helmholtz had described in his previous paper. Therefore, it was necessary to isolate the portion of mixed colour being observed from all the others surrounding it. This he did by the simple expediency of looking through the telescope from about one or two feet away rather than the usual manner of placing the eye directly against the eyepiece. The field of view was quite small and the colour being viewed was free from the disturbing effects of the colours closeby. The cross hairs in the telescope marked the area being viewed and if these were not visible from the distance to the eyepiece they could be made so by looking through a weak concave lens held in front of the eye. [2, p254] This had the effect of making the observer long sighted!

One final step completed this careful arrangement. A dark moveable screen with a small hole was placed at the viewing distance. Actual observations were made from behind the screen looking through the hole to the telescope eyepiece. With the very restricted field the hole served to mark the exact colour combination in view while it eliminated the need to move one's head around to locate the colour if viewing was interrupted. In order to know what simple colours were producing the mixture a second prism could be held between the hole and the eye to resolve them. An alternative solution of that problem was to have a second person close one of the slits, eliminating one of the two colours. Therefore, between the cross hairs aligned on the colours or on Fraunhofer lines and the ability to resolve the compound colour, the possibility of mistaking the two simple colours of a mixture was eliminated. The result of all these arrangements was that it was 'possible to obtain the total combination of every two of the simple prismatic rays in all degrees of relative strength, and to observe them undisturbed by the presence of other colours.' [2, p525]

The main results of his observations Helmholtz summarised in Table 5.1. [2, p534] He placed there five spectral colours which he thought were sufficiently accurate to represent the spectral mixtures in their major variations, although we will note that purple is not a spectral colour. The five colours are listed at the top horizontally and at the left vertically, a mixture of

Table 5.1. Helmholtz's mixtures of spectral colours

	Violet	Blue	Green	Yellow	Red
Red	Purple	Rose	Dull yellow	Orange	Red
Yellow	Rose	White	Yellow-green	Yellow	
Green	Pale blue	Blue-green	Green		
Blue	Indigo	Blue			
Violet	Violet				

any two of them being their intersection. The most surprising thing to note in terms of the unquestioned acceptance of red, yellow and blue primaries is that of all these spectral combinations only yellow and blue produced white. (Helmholtz called this blue an indigo-blue.) But according to those who held red, yellow and blue as primaries, a mixture of yellow and blue should have given green, not white. We have here at last a definitive experimental observation which refuted that belief, which had been based, of course, on the mixture of pigments rather than on mixtures of lights such as Helmholtz had effected. Moreover, in all discussions by previous investigators on compounding white from different colours, this was the first experimental demonstration of the fact since Newton. We shall return to this point at the end of the chapter.

The data above indicate that yellow is a combination of red and green and that yellow and blue give white. The implication is that red, green and blue should yield white when mixed. If so, it would vindicate that age-old belief that three primary colours are required to make white—assuming, of course, one followed Newton's opinion on this. Heretofore, all theories of primary colours had simply assumed this was so.

In the preceding experiments, Helmholtz had used two slits giving binary colour combinations. Now the arrangement was modified so that three slits produced overlapping spectra. In a black screen Helmholtz cut two parallel slits inclined at 45° to the horizontal. They were a distance apart so that 'when observed through the prism at the ordinary distance, the violet of one fell upon the red of the other.' [2, p527] Hence the overlapping strip was purple. Now a third slit at right angles to the other two was cut between them so that the three slits together looked approximately like a Z. Green from this spectrum was allowed to overlap the purple from the other slits, doing so at right angles. Thus there were various colour triples formed all tending toward white. As before, the relative brightness could be changed. From this arrangement together with the previous results Helmholtz concluded,

> with respect to the combination of three simple colours, we may conclude that white can only be the result when rays from the three different sections of the spectrum are suitably united. At least it cannot be supposed, although all possible combinations cannot, of course, be exhausted by experiment, that the yellow or yellowish colours, for example, which are derived from the red and green sections, can, by the addition of one or more colours contained in these sections, red, yellow,

or green, pass over into white. This remark is also applicable to the mixtures of the green and violet, as also to those of the red and violet sections. We may, on the contrary, succeed in obtaining white from various combinations of three colours taken simultaneously from the three sections. [2, p527]

And we have here the first experimental evidence that three primary colours (of light) do indeed make white. That long standing assumption had at last been verified.

In addition to testing the assumption about primary colours, Helmholtz attempted to find natural complementary spectral colours other than yellow and indigo-blue. He was able to do so however only by arranging the relative intensities to obtain a pure, untinted white. His results were:

simple red and compound dull blue-green
simple green and compound purple-red
simple violet and compound dull yellow.

But the V-slit arrangement had indicated that yellow and indigo-blue were complementary colours with no adjustment necessary to their intensities as compared to these pairs produced by adjusting the intensities. Helmholtz concluded therefore that only two colours in the spectrum were properly complementary, yellow and indigo-blue. (This, however, was to be proved theoretically wrong by Hermann Günther Grassmann one year later while additional experiments by Helmholtz inspired by Grassmann's work confirmed the correctness of Grassmann's ideas.)

It seems all discussions of colour in the spectrum harken back to Newton. Every investigator in this history had found authority for his opinions in the work of the Master. Even Helmholtz felt bound by this tradition and called attention to the fact that the few combinations of spectral colours which Newton did mention all confirmed his own results. [2, p528] Unfortunately in the passage of interest to Helmholtz there was another statement which must have given comfort to the proponents of the red, yellow and blue primaries. In describing his binary colour combinations, Newton stated,

> And after the same manner other neighbouring homogeneal Colours may compound new Colours, like the intermediate homogeneal ones, as yellow and green, the colour between them both, and afterwards, if blue be added, there will be made a green the middle Colour of the three which enter the Composition. [4]

The point being made by Newton was that each compound colour mentioned was always the intermediate between the two simple colours. Thus it is easy to see how people could have followed the pigment analogue into the spectrum insisting that yellow + blue = green, since green is intermediate between these two in the spectrum. However, intermediates need not be the only possible results of colour mixing for Helmholtz had demonstrated that yellow + blue = white, which properly speaking is not a colour. Moreover, the production of purple could not be called an intermediate colour between red and violet since it does not exist in the spectrum. As mentioned earlier Helmholtz's experiments with spectral colours are really the first of their kind since Newton. No mention of experiments combining spectral colours in the eighteenth century was made by Priestley in his *History and Present State of Discoveries Relating to Vision, Light and Colours*, [5] and the previous chapters have covered the nineteenth century to date. Not only was the list of primary colours based on pigment mixtures taken for granted as axiomatically true, but the authority of Newton further strengthened the argument for it with apparent experimental evidence. Helmholtz's experiments are then all the more remarkable. The result of combining yellow and blue refuted an argument on colour mixing by Newton and, as Helmholtz said, it contradicted 'in the most decided manner the experience of painters during the last thousand years.' [2, p529] It is perhaps no light thing to do away with such a tradition, but the experimental evidence could not be denied.

The result raised a problem of explanation. For it is one thing to find evidence contradicting ancient opinion but another to bring such new observations within the realm of current knowledge. Therefore, the next steps in Helmholtz's elucidation of colour mixing were twofold: (i) explain why pigment mixing does not give the same result as light mixing, and (ii) reconcile the facts from both areas in terms of known optical facts.

The question was essentially this: What is the physical mechanism by which coloured substances act on light? The first point is that some pigment substances may be transparent (or at least translucent) while other pigments may be opaque. For transparent pigments (the only case considered by Helmholtz) the light transmission varies for different wavelengths. When light is incident on a coloured surface or powder some of the light is immediately reflected from the uppermost layer, the surface, as white light. But some of the light must pass beyond the surface into the layers below where most of the colour will be absorbed. Only the colour which can penetrate to some depth and be reflected from the interior of the pigment materials, again without absorption, reaches the eye. It is this remaining light which gives the pigment its characteristic colour. The process is subtractive in the sense that pigments remove most of the light from the incident beam reflecting back only the colour to which they are transparent. For confirmation of this fact, a glass plate reflects only 1/25 of the incident light from its surface while two such plates reflect 1/13 of the incident light and in a similar manner many plates reflect almost all the incident light. [2, p529] It follows for powdered glass, rather than plate glass, that most of the light is reflected from the interior layers and very little from the upper surface. And so for powdered pigments analogously, most of the reflected light is from below the surface and then mainly the colour to which the pigment is transparent.

What happens then if we mix two coloured powders, say yellow and blue? Blue substances are relatively transparent to green, blue and violet but stop red, orange and yellow; yellow substances are relatively transparent to red, yellow and green but stop blue and violet. When light is incident on such a mixture, some of it is reflected from the surface. The blue powder reflects blue, the yellow powder reflects yellow and the two surface reflections combine to give white. But as noted, the fraction of light reflected from the surface is small, most of the reflection taking place from the interior layers below the surface. The only kind of light which can penetrate both pigments and be reflected from the interior without complete absorption is green. For it is only green to which both powders are transparent. Therefore, a yellow and blue pigment mixture should indeed give green according to the laws of transparency. Moreover, since that explanation postulates that most of the incident light is absorbed it also explains why mixtures of pigments are darker than either of the colours separately. It was this understanding that Forbes had approached but never reached. The process of mixing pigments is inherently a subtractive one while the mixture of lights is inherently an additive one, hence Helmholtz argued that the results of colour mixing from one process cannot furnish any information regarding the other. [2, p530] The process of pigment mixing was at last seen to fall within the known laws of physics while the process of light mixing was seen to be governed by heretofore unrecognised laws. A search for the foundations of the latter occupied later investigators and is the subject of later chapters. (It is worth pointing out that opaque pigments, not discussed by Helmholtz, also mix colours subtractively so that his basic conclusion with these materials is still valid. However, their action is different being essentially a surface effect of absorption and light scattering.)

Although the experimental method described above was a powerful one Helmholtz sought confirmation of his results by two other methods. Of these, one was the spinning disc, while the other is shown in Figure 5.2. [6] Suffice it to say that results obtained by both methods agreed with those of mixing prismatic colours.

Having produced and even classified many combinations of coloured lights, Helmholtz turned to that ancient question of primary colours.

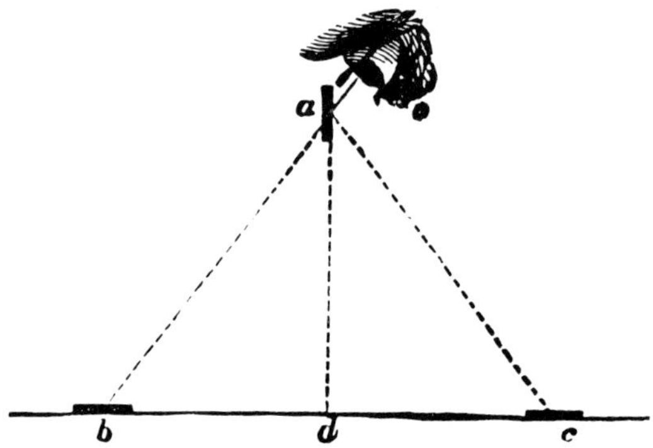

Figure 5.2 Helmholtz's wafer method of producing additive colour combinations. The observer sees coloured square *b* directly and coloured square *c* by reflection. The two separate colours of light entering the eye give an additive colour combination.

However, he restricted himself to imitating the spectral colours rather than all possible hues in nature. He could produce a compound orange from red and yellow which was essentially indistinguishable from the simple spectral orange. Similarly he produced a compound indigo from blue and violet indistinguishable from the simple spectral indigo. But a compound red of orange and violet was a poor imitation of spectral red and he could not produce a good imitation of spectral red from mixtures of any spectral colours. Spectral violet too was poorly imitated by a mixture of indigo-blue and a little red. Imitations of spectral yellow and green from their respective neighbours were similarly not good although better than the compound red and violet. Thus if the problem of primary colours was defined as imitating the spectral colours with the least possible number of simple colours, then Helmholtz stated that five colours were necessary: red, yellow, green, blue and violet. [2, p532] These are, of course, the colours in Table 5.1. Even on this choice he was doubtful, wondering whether differences he had not been able to detect might not be detectable with a larger field of view. [2, p532] This surprising choice was, of course, based on a restricted definition of the problem: that of imitating only *spectral* colours. His predecessors had never made such a distinction; indeed their object was to imitate 'nature's' colours, in which category they included spectral colours as well as pigmented colours. So Helmholtz was really looking at a modified problem. It was nevertheless an uncomfortable choice for the idea that the primary colours are three rather than five had long been entrenched; Helmholtz therefore tried to bring his observations within the concept of three primary colours. Towards this end he stated:

> If, however, we wish to limit ourselves to three colours, it would be best to choose the three simple ones which admit of the least perfect imitation, namely red, green, and violet; we should then obtain a yellow and blue, which, compared with the colours of our pigments, would appear saturated, but which would not bear comparison with the yellow and blue of the spectrum. [2, pp532–3]

Although this sop was thrown to the old notion of three primary colours, Helmholtz thought that Young's trichromatic idea consisting of three basic sensations would have to be abandoned. For he said:

> If the sensation of yellow by the yellow rays of the spectrum were due to the fact that by them the sensations of red and green were simultaneously excited, and both working together produced yellow, exactly the same sensation must be excited by the simultaneous action of the red and green rays; nevertheless by the latter we can never obtain so bright and vivid a yellow as that produced by the yellow rays. [2, p553]

Applying similar reasoning to the blue and violet, colour sensations or colour receivers in the eye in the sense of Young would require the five colours mentioned above. The three colour theory would be sufficient only to classify the non-spectral colours of natural bodies in the sense of Mayer and the colour organisers. [2, p533] Helmholtz was not proposing a new hypothesis in place of Young's but only remarking that on the basis of his own experiments it seemed inadequate. Later, however, in his *Physiological Optics* (1867) he retained Young's theory and even attempted to extend its applicability as a means of understanding the physiology of vision. [7, pp144–5]

DISCUSSION OF HELMHOLTZ'S WORK

The history of science contains numerous cases of assumptions so evident as to be unrecognised and which have ultimately led to a conceptual crisis. The stationary earth, the classical principle of relativity, the luminiferous ether and the commutation rule in algebra are a few examples. In each case the tacit assumption worked well enough to allow continuous evolution of the subject until anomalous results appeared. Then the consequences of maintaining the old persuasion caused confusion rather than enlightenment and new insights were necessary to reveal the error of the 'self-evident truths'. To this list must be added the present one dealing with mixtures of lights and pigments. These, as noted, were always considered equivalent.

The events in Chapters 1–4 can be viewed in the context of this one false analogy. Young, for example, applied it to the eye, and arrived at a potentially useful concept of colour perception. Brewster was less lucky in his application, and in fact it is doubtful if Brewster's theory could have emerged had he not been subject to the false assumption that red, yellow and blue are primary colours for vision and that $Y + B = G$. As quoted in Chapter 2:

> This glass absorbed the blue rays, which, when mixed with the yellow, made green, and the yellow rays, which when mixed with the red, made orange; and by insulating the yellow and red, it thus affected a perfect analysis of the compound green and the compound orange. [8]

This 'result' seems to have especially impressed Brewster. Freed from its dependent clauses and without the interpretive remarks at the end it reduces to

> This glass absorbed the blue rays . . . and the yellow rays . . .

The clauses omitted contain Brewster's tacit assumptions about light mentioned above plus the additional assumption that $Y + R = O$, which is correct. With respect to this paragraph we have already said that what remained for Brewster to see was only red and violet. Rather than state positively what in fact he saw, Brewster reported the colours he thought the glass absorbed, a basically negative statement. The colours that were missing were blue and yellow which was true enough but it was actually incomplete as an observation because the glass absorbed orange and green as well. However, because of his assumptions that these colours must be compound colours the second part of the paragraph, which is mainly interpretative, is nothing but an assumption, and a wrong one. Had he not assumed orange and green to be compound colours and had he not further assumed the compounding in terms of red, yellow and blue, he would have had no explanation for the absence of green. And that means he could not have applied these primaries to form the other spectral colours. Had he insisted on the same form of explanation for the spectrum he would at least have had to modify his choice of primary colours and the combinations obtainable from them in order to fit his observations. But the standard assumption was red, yellow and blue. It is therefore a remarkable coincidence that Brewster's absorption observations so neatly fitted a ready-made pattern of primary colours and colour combinations. There can hardly be any doubt that the ease with which they fitted together encouraged him in his reasoning toward the triple spectrum. The entire theory in effect was based on a false analogy with pigments whose extreme development became an anomaly in physics violating well developed ideas. Therefore it had the effect of forcing a reappraisal and redefinition of the question of compound and primary colours. In that sense it ultimately served a useful purpose.

The analogy has been traced back at least to Newton. Yet if Newton made this mistake, it is those who came after him who must be faulted. It has been said that 'an idea is not responsible for the people who believe in it.' It appears that in later years, especially in the later eighteenth and early nineteenth centuries, no one actually tried to reproduce his results. They were taken for granted and he became in some respects the authority to whom appeals were made without critical analysis. Young is indicative of this attitude not only invoking Newton's words but even stating that no new experiments were needed for his theory of light. We must in fact conclude that so far as the theory of colour and colour perception was concerned, until the mid-nineteenth century, that seemed to be so.

Yet considering the difficulties surrounding an understanding of compound colours, it was Helmholtz who finally analysed the basis of that confusion. His experiments against Brewster and those described in this chapter can be viewed very much as a return to Newton. They were finally a return to fundamentals aimed at repeating what were essentially Newton's experiments in order to determine for himself what combinations could be made from spectral colours. That was, after all, almost the starting point for Newton. But at the same time it was necessary to perform them in a highly refined version; as in his approach to Brewster's experiments, he repeated them with the errors removed. For Newton's observations of compound spectral colours depended upon moving a white card within the field of colours formed from a prism and focused by a lens, or moving a comb-like device in front of the pencil of colours to produce delimiting slits for certain colours. Certainly these methods did not meet mid-nineteenth century standards.

Helmholtz required results from a stationary arrangement together with the ability to look closely at binary combinations without interference from other colours. This was never achieved by Newton. It was part of Helmholtz's genius to see that he could produce both results from the simple expedient of the V-slit coupled with a viewing telescope. (He also had an achromatic telescope, a relatively new invention.)

As the first such experiments since Newton's time they were done in a highly convincing and repeatable manner. Helmholtz demonstrated that the heretofore tacit assumption that yellow and blue light make green was wrong and with that demonstration he forced an abandonment of the pigment analogy upon which it was based. Together with his observational evidence he also showed theoretically that that analogy had been a false one because in fact mixtures of pigments and mixtures of light obey quite different laws. For all those years, no analogy had ever really been possible.

Considering these events, it seems that this work can well be judged as a small scientific revolution. Small or not, however, it had important ramifications for later nineteenth and twentieth century research in colour perception, colour technology and illumination. Helmholtz almost singlehandedly achieved a physical understanding of compound colours. In this achievement he not only closed one era but opened another, one in which he himself, of course, played an active role. After this work, necessarily qualitative, the ground was prepared for quantitative and mathematical investigations by Grassmann, Maxwell and others. It is significant that the work of Grassmann and Maxwell followed soon after with consequences of the new model.

NOTES

1 Plateau, Joseph A F *Dissertation sur quelques propriétés des impressions produites par la lumière sur l'organ de la vu* (English translation 1924 James P C Southall (ed) vol II (New York: Optical Society of America)). The original article remarked on by Helmholtz in his *Physiological Optics*

2 von Helmholtz, Hermann 1852 On the theory of compound colours *Ann. Phys., Lpz.* **9**. (English translation 1852 *Phil. Mag.* **IV**.) References here are to the English version

3 The drawing is based on one in Helmholtz's *Physiological Optics* (see [7])

4 Newton, Isaac 1730 *Opticks* 4th edn (London). Reprinted 1952 (New York: Dover) p133

5 Priestley, Joseph 1772 *History and Present State of Discoveries Relating to Vision, Light and Colours* (London)

6 The method is discussed by Helmholtz in [2]. Figure 5.2 is based on a similar one in Helmholtz's *Physiological Optics* [7]

7 von Helmholtz, Hermann 1860 *Physiological Optics* 1st edn. The third edition of 1911, from which the English translation was made, was based on the first edition

8 Brewster, David 1831 On a new analysis of solar light *Edinb. J. Sci.* **V** also published in 1834 *Trans. R. Soc. Edinb.* **XIII**

6
Hermann Günther Grassmann: Towards a Quantitative Theory of Colour

Descriptions of compound spectral colours had been brought to a new standard by Helmholtz in 1852. But in showing how to proceed in this science he had also reached two surprising conclusions. His opinion that only yellow and indigo-blue were *natural* complementary colours was at variance with Newton's colour circle which implied many pairs of complementary colours in the spectrum. Also he thought that the minimum number of primary colours was five which contradicted long-held opinion. While he was willing to accept these conclusions against entrenched opinion, Hermann Günther Grassmann, one of history's neglected geniuses, would accept neither. Convinced that Newton's circular colour classification was ultimately correct Grassmann set out to justify it mathematically and to correct the errors he saw in Helmholtz's experiments. [1, p254] To this end Grassmann proposed the following questions: (i) are additive colour mixtures a manifestation of only three primary colours or do more than three variables enter the phenomenon of colour perception; (ii) can it be shown theoretically that the spectrum contains many natural pairs of complementary colours; and (iii) what is the mathematical form by which we can quantify the colour facts summarised in Newton's colour circle? The answer to the first question came from a totally unexpected direction while the answers to the other two employed Grassmann's own earlier work on vector analysis together with ideas borrowed from photometry to produce profound new physical insights into colour mixtures.

THE VARIABLES OF COLOUR VISION

Since any physico-mathematical theory must first define its variables, Grassmann at the outset determined 'to analyse the impression of colour of which the eye is capable into its elements.' [1, p254] It was necessary to determine clearly what physical parameters are distinguishable by the eye. Considering white light, which he called colourless, Grassmann pointed out that the eye can distinguish only levels of varying brightness. Similarly for a given homogeneous or monochromatic colour, the eye can distinguish only greater or lesser brightness. Thus one variable of colour vision was that of brightness which is measurable in terms of photometry.

Another variable of colour vision exists in the colour itself. For a given level of brightness, the eye can distinguish various homogeneous colours, which are measurable in terms of their wavelengths. (Grassmann, however, spoke of the period of each vibration, a correct but somewhat obscure definition of colour.) This variation of colour Grassmann noted is recognised in ordinary speech by the word 'tint', but is now more properly called 'hue'.

In short, Grassmann had defined two observational variables for pure colours: hue and brightness. Now he proposed that even if both the hue and brightness of a given homogeneous colour remain constant, there is yet another means of varying colour impressions. This he pointed out is done simply by mixing white (colourless) light together with any given homogeneous colour. Then ostensibly the colour must retain its original hue and original brightness but its appearance will become weakened and pale. Employing a linguistic argument, Grassmann noted that our speech already recognised this phenomenon:

> ordinary language is rich in terms by which this difference is characterised; the expressions, intense, deep, pale, dull, whitish, saturated, which are added to the names of colours, may serve as instances. [1, p255]

But Grassmann thought it was necessary to replace these popular expressions with a scientific term signifying the physical act of mixing white light with a pure colour. This third variable he called the 'brightness of the intermixed white', [1, p255] [2] a term denoting the property of a colour of constant hue as being more or less white. It is today called 'saturation', or purity. So-called deep colours such as spectral colours are literally saturated with colour while pale colours such as pastels containing much intermixed white are desaturated. (Desaturation was noted by Newton in his circle by the distance of a colour mixture from the circumference towards the centre.) It must be noted, however, that Grassmann's assumption about mixing white light with a given colour is not strictly correct. Admixture of white with a homogeneous colour will in fact shift its apparent hue slightly toward yellow in addition to altering its saturation, but this fact was only noted in 1910 by W deW Abney with observational methods not available to Grassmann. [3] For Grassmann it was a reasonable, essentially valid assumption, which served his immediate purposes well and it was ultimately to show by argumentation that any given colour can be characterised by only three visual elements: hue, brightness and saturation

To illustrate that these variables are actually experimental observables, which can specify any colour, Grassmann suggested a procedure by which they could, in principle, be measured. Two similar white screens hinged together are placed on a table as shown in Figure 6.1. The screens are to face the

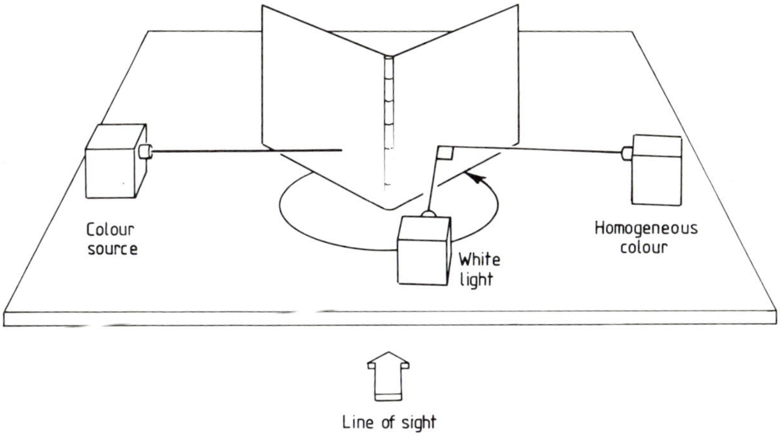

Figure 6.1 Schematic illustration of Grassmann's thought experiment. The visual equality of colours can be measured in principle when the hue and luminance of the two screens are equal.

outside of the angle formed at the hinge. A divided circle between the screens measures the angle between them. Let an arbitrary (mixed) coloured light from a source on the table be incident at any oblique angle on one screen, say the left. The other screen is illuminated with two lights, from separate sources. One light is homogeneous of the same apparent hue as that on the left screen, the other light is white so the right screen now has a mixture of homogeneous light and intermixed white. The two lights on the right may be incident on the screen at any oblique angle but must be perpendicular to each other. Clearly the colour impressions received from the two separate screens need not be identical. Now rotating the right screen varies the brightness ratio between the reflected homogeneous light and white light as seen by an observer looking towards the screens. Rotating the left screen also changes the brightness of light reflected from there. Providing the brightness on the right screen is not too great we should be able to find positions for each 'in which both will produce the same impression upon an eye observing them both at the same time.' [1, p256] Since the three variables consist of two measurements of brightness (the homogeneous light and the intermixed white) and one of hue (wavelength) this arrangement can, in principle, measure the two brightness variables in terms of the angle formed by the screens. The third variable of hue would be predetermined by known optical means. The method is a photometric one, using the eye as a null detector device; for although the eye cannot judge the ratio of illumination between the two screens it can quite accurately judge their equality. There is a difference however between Grassmann's concept and the usual methods of photometry then in use. In the latter white lights were used as often as possible so that observation was reduced to matching brightness only. Here, however, one must not only match brightness but hue as well, perfectly possible in principle but much more difficult in practice. Nevertheless the principle involved here is now well established in modern colorimetry and has been stated as follows: 'every mixture of lights can be matched by a definite spectral light or a definite purple mixture (i.e. red and violet) which is mixed with a definite amount of white light.' [4] Stated otherwise, two lights of the same colour specifications (hue, saturation, and brightness) produce the same colour impression in mixtures regardless of their actual spectral composition. This is the most fundamental principle of modern colorimetry for it allows one to experiment with coloured light solely on the basis of colour sensation without regard to specific spectral content.

These statements together with the thought experiment were ultimately plausibility arguments for the elements of colour vision. Nevertheless Grassmann regarded them as empirically well proven for he pointed out that no observer had as yet been able to mention any *other* means than these three variables by which the sensation of colour might be altered. Employing yet another linguistic argument in favour of this idea. Grassmann said,

> And moreover, in ordinary language, we only recognise these three elements in the description of this phenomenon, so that we may affirm with certainty that hitherto only these three elements have been observed in the impression of colours. [1, p256]

The evidence for any number of variables of colour vision other than three was conspicuous by its absence. For example painters usually describe colour variations only in terms of hue, tint, and shade. In Grassmann's nomenclature the artist's 'hue' is equivalent to Grassmann's tint (wavelength); the artist's 'tint' is Grassmann's brightness of intermixed white (saturation); the term 'shade' refers simply to Grassmann's brightness of the light.

Grassmann had now answered his first question but in a most unexpected form. Whereas everyone else had thought only of three primary colours, Grassmann by persuasive argument had shown that colour perception was indeed a three-fold process but dependent on quite different physical parameters. But in preserving the three-fold nature of colour vision which Helmholtz was willing to concede he simultaneously challenged the cherished notion of primary colours. It was Maxwell who would resolve this apparent conflict two years later.

COLOUR TRANSITIONS AND COMPLEMENTARY COLOURS

Grassmann next proposed the hypothesis 'that if one of two mingling lights be continuously altered (whilst the other remains unchanged), the impression of the mixed light is also continuously changed.' [1, p256] This may seem fairly obvious, yet the fact is that this was the first explicit statement of what had previously been only an intuitive idea. For purposes of a quantitative theory it was necessary to state clearly the suppositions on which it was based.

By this principle, a colour impression can be changed by altering the brightness of the light. By varying either the brightness of a given colour, or of the intermixed white, or both, one could change a given colour sensation into quite a different impression. For example, if the brightness of the colour itself were decreased to zero the light would become colourless since only the intermixed white would remain on the screen. At the same time one could increase from zero the brightness of quite another colour. Thus one colour would fade out while the other faded in giving the impression that the first colour had changed into the second by a continuous transition. Grassmann said of this:

> By the continuous diminution of the brightness to zero, a hue may gradually pass over into another completely separated from it, if the brightness of the latter increase continuously from nothing upwards. [1, p256]

Here the two different hues have remained constant. But colour impressions could also be altered by a continuous change of wavelength. In the spectrum there is a continuous change of colour from red to violet. But Newton had arranged the spectrum in a circle (rather than a straight line) mentioning the existence of purple transitions between red and violet at opposite ends of the spectrum. Newton had even considered purple simply a 'compounded violet' stating with respect to his colour circle that

> if the point Z fall in or near the line OD, the main ingredients being the red and violet, the Colour compounded shall not be of any of the prismatick Colours, but a purple, inclining to red or violet, accordingly as the point Z lieth on the side of the line DO towards E or towards C, and in general the compounded violet is more bright and fiery than the uncompounded. [5]

One can argue then that there exists a series of colour transitions between red and violet yielding various purple sensations. Indeed one must argue this if Newton's circle is to be maintained as a basis of colour theory as was Grassmann's objective. Thus Grassmann argued further that there is a continuity of change between violet and red through the purples so that those transitions between violet and red must also be included in the idea of continuous change of hue. Grassmann defined two types of colour transitions. Referring to Figure 6.2, transitions of hue from red through orange, yellow, green, blue, violet and purple back to red he called positive transitions. Transitions in the reverse direction he called negative. [1, p257]

There were for Grassmann basically three different ways by which coloured light A could continuously pass into coloured light B. These were changes of hue in the positive direction, changes of hue in the negative direction, or changes of hue by 'the light becoming colourless once or several times during its transition.' [1, p257] By whatever route a colour transition takes place, however, Grassmann noted that all transitions must be regarded as falling within the mathematical idea of continuity including transitions between red and violet through purple. Arguing again from observation, Grassmann stated:

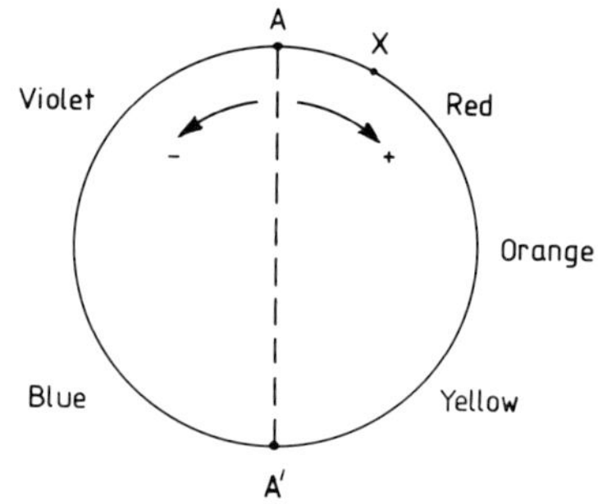

Figure 6.2 Colour transitions around the circle to prove the existence of many pairs of spectral complementary colours.

> the principle of continuous transition . . . must be regarded as perfectly established by experience, as a sudden jump in the phenomena would be apparent even in the most crude observations, and such a jump has not as yet been observed. [1, p257]

There were no quantum jumps in colour transitions!

Grassmann was able to show that Helmholtz's observation of a single complementary pair of colours must be wrong. He proved the following proposition:

> To every colour belongs another homogenous colour, which when mixed with it, gives colourless light. [1, p257]

The nature of the proof was 'reductio ad absurdum'. Taking the given colour to be A, Grassmann assumed there was no other homogeneous colour which together with A would give white light. And then from this he reached a contradiction of the negative assumption thus showing it to be wrong. Therefore, the only other possibility, that the colour A did have a complementary colour must be correct. Since the colour A was arbitrary, the proposition was then proved.

In Figure 6.2, A is any given colour. Assume no other homogeneous colour gives white light when mixed with A. Take a second colour, X of any wavelength. Let the second colour have brightness Y. We now mix A and X together. Keeping the wavelength of X constant, we increase its brightness, Y, from zero. We shall reach a brightness Y (of X) where the brightness of A will be negligibly small compared to Y. In effect, the colour A will disappear by comparison with X. But since we have assumed the two colours cannot give white the change must be one of hue. Therefore the transition has been a continuous one in hue from A to X. But transitions in hue are either in the negative direction along the circle, or in the positive direction. Grassmann now investigated the consequences of the two different modes.

Suppose hue X is infinitesimally different from A but on the side toward positive transitions. Then the transition from A to X is positive. This is required by the condition that colour transitions be continuous, and we can argue this as follows.

Mixing A with a new brightness level of A, A_Y, still gives the same colour A. Only the brightness of A has changed. But by increasing the brightness of X sufficiently the colour transition of A to X occurs. If X were on the positive side of A, but the transition took place in the *negative* direction, there would have to be transitions through all the other colours around the circle back to A. Since only two colours are involved, and they are infinitesimally close, this means colours besides those differing infinitesimally from A will be produced. This is contrary to observation and also violates Grassmann's requirement for mathematical continuity. Therefore the change from A to X is in the positive direction if X lies on the positive side of A and infinitesimally close to A. By the same reasoning we can say if X lies on the negative side of A, infinitesimally close to A, the transition is negative.

Suppose now in Figure 6.2 we have a hue A and opposite it the hue A′. It is desired that X changes continuously from A in the positive direction passing through all the colours back to A. Each transition is made by changing the brightness, Y. For half the journey around the circle X lies to the positive side of A so the transitions are positive. But for the other half of the journey X lies to the negative side of A so the transitions must be negative. Thus in bringing X around the circle back to A, the sign of the transition changes.

Suppose A′ is the colour where the sign changes. Then transitions are positive just before X reaches A′ and transitions are negative just after passing through A′. Suppose further that X passes continuously through A′ as required by the hypothesis of continuous colour changes. Since the transitions are being made only by changes of brightness of X, and since the hue must change continuously, this means all the colours resulting from an increase of brightness of X lie extremely close together on *both* sides of A′. But this is impossible. An increase of brightness of X can give either a positive or a negative transition, but not both. And it is here that the negative assumption is contradicted. The implication of this result is that to reach A′ from A one must pass along the diameter of the circle instead of the perimeter. But in Newton's circle of colours the diameter passes through white so a passage from A to A′ necessarily requires a transition through colourlessness. And such a transition was specifically allowed by Grassmann in the three modes of transition mentioned above. Hence the assumption that no other colour existed which when mixed with A would yield white led us to attempt to make continuous transitions around the circle. The attempt failed at A′ where the transitions change sign discontinuously. And so A and A′ are complementary with combination at the centre of the circle, or white. Therefore, Grassmann had shown that by assuming colour A has no complementary colour it is impossible to make continuous transitions from A to all the other colours. And it follows that the original proposition must be correct; every colour has a homogeneous colour which, when mixed with it, furnishes white. This colour lies opposite the given colour along a diameter of the circle. This is the very important consequence of the 'obvious' statement that if one of two mixed lights is continuously altered the visual impression also continuously changes. Unfortunately, however, this proposition is not entirely correct and led Grassmann to an error as we shall shortly see.

The foregoing proved the existence of many pairs of complementary colours assuming the validity of Newton's colour circle. It did not show specifically what those complementary pairs were. Therefore, Grassmann's next goal was to utilise Helmholtz's observations to predict additional spectral complementary colours.

It is not clear from Newton's description whether he had divided his circle in proportion to the width of the colour bands in the spectrum as he viewed them. If so, his division in 1853 was no longer valid considering the work of Fraunhofer. So to determine the complementary pairs by Newton's colour circle Grassmann revised Newton's division of the circle according to Fraunhofer's description of the spectrum. Considering the differences in dispersive power between Newton's and Fraunhofer's prisms, Grassmann thought it was reasonable to assume that the boundaries of Newton's circle would coincide with Fraunhofer's B and H lines. Newton's first (or far) red would coincide with Fraunhofer's line B which lies toward the far red; Newton's last violet would coincide with Fraunhofer's line H which lies toward the far violet. Then he marked off various well known Fraunhofer lines on the circle and from these reference marks layed off the colour divisions according to Newton's and Fraunhofer's measurements. Grassmann's revision of Newton's circle is shown in Figure 6.3. The inner circle of Grassmann's figure shows the positions of various Fraunhofer lines drawn as short lines and marked with capital letters. Then the limits of colours of the circle are drawn to the outer circle for clarity. As expected the division of the circle is similar to Newton's but there are adjustments to the colour sectors, while Newton's line OD becomes the one marked here with Fraunhofer's lines B and H. [6]

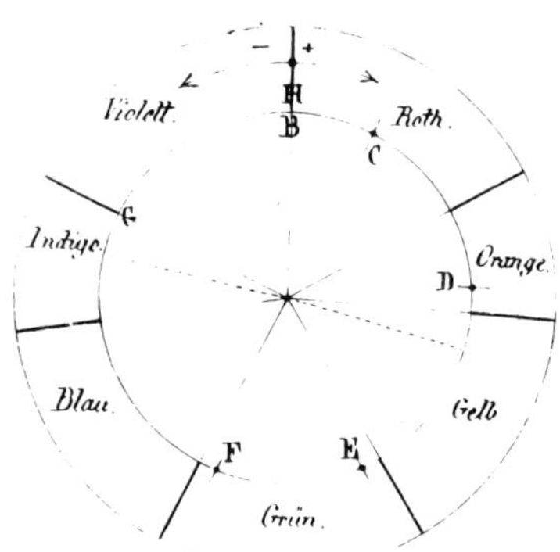

Figure 6.3 Grassmann's revision of Newton's colour circle.

Grassmann was now able to locate Helmholtz's yellow and indigo-blue complementary pair. Helmholtz had defined the yellow as 'lying between the

Fraunhofer lines D, E and about three times more distant from E than from D.' [7, p526] This fixed the position of yellow on the circle, and, of course, the indigo-blue must be opposite it along a diameter. These are shown in Grassmann's diagram on the dotted diameter. Now combinations of red and green give rise to positive transitions. When red predominates, the mixture is toward the red; when green predominates, the mixture moves toward the green. But according to Helmholtz roughly equal amounts of red and green gave a pale desaturated yellow [7, p525] which indicates a transition toward the centre of the circle. This is not along a diameter, however, as white is not produced, but must be on a chord connecting red and green. On the other hand, red mixed with violet or with blue or indigo must have given negative transitions, and red and green-blue for Helmholtz gave a flesh colour. This again is essentially a desaturated colour being, according to Grassmann, 'nothing but a red mixed with much white.' [1, p260] Now there can be no transition from the flesh colour along the perimeter to the blue-green; the only transition from the flesh colour to blue-green is by 'weakening the red until it entirely disappears before the white.' [1, p260] The implication is that flesh colour also lies along a chord (between red and blue-green) and that the diameter between red and its complementary colour lies between the two chords described. Thus to go from red to blue-green must be via a colourless light (white) transition through the centre of the circle. Grassmann therefore concluded that the complementary colour of red must lie between green and azure-green and be a form of bluish-green. [1, p260] By similar analysis he produced the following table of complementary colours:

yellow	yellowish-green	green	bluish-green	azure
indigo	violet	purple	red	orange

He had now completed his second objective which was to show that there must be other natural pairs of complementary colours (in the spectrum) besides the sole example reported by Helmholtz. These theoretical results were fully confirmed and extended by Helmholtz himself less than six months later.

As stated before, Grassmann's proposition is not correct as given. There the strict requirement is that every spectral colour must have a *spectral* complementary colour, but his own table of complementaries includes purple which is not a spectral colour. Helmholtz's later confirming results notwithstanding, the fact is that every spectral colour does not have a spectral complement in the sense required by Grassmann's proposition. There is a range of colours in the greens from about 484 nm to about 570 nm for which spectral complementaries do not exist. Their 'complements' are found in the purples. Including purple in his table of complementaries is consistent with his idea of purple as a transitional colour between red and violet; but it is inconsistent with the stated proposition itself that every spectral colour has a spectral complement. To be consistent with his idea of mathematical continuity through the purples, Grassmann's colour circle should have explicitly contained purple between violet and red. Then his proposition could be understood, as indeed it is today, to say that every spectral colour has a complementary colour although the complements of the greens lie in the purples which are not found in the spectrum. This was also Grassmann's own understanding of his proposition as shown in his own illustration (Figure 6.5).

The apparent contradiction between Grassmann's proposition and his own correct results lies in his views of red, violet and purple. In a brief discussion of the extreme ends of the spectrum Grassmann says of red, 'this outermost colour of the spectrum, which may be equally well regarded as extreme red or extreme violet, I will call purple.' [1, p255] For Grassmann the red and violet extremes looked alike or produced equivalent visual sensations which is not so. But just before this remark Grassmann indirectly refers to purple as being observable in the spectrum. Citing a previous article in Poggendorff's *Annalen*, Grassmann stated that purple could be observed in the spectrum 'under remarkably favourable circumstances (on a bright summer's noon).' [1, p255] [8] Amazingly, there was still some doubt, even then, on this point since some observers claimed to see purple under these conditions while others denied it to be so. (Probably they observed violet which was then and is still called colloquially, but incorrectly, purple.) Even Helmholtz in confirming Grassmann's results shortly afterwards, considered the question still outstanding; as one of his new observations he demonstrated the fallacy of this belief. [9, pp10–12] Grassmann, however, not being an experimentalist, simply accepted his authority as reliable even though those observations dated as far back as 1799. Regarding purple then as a spectral colour, Grassmann meant in the quotation above that the 'outermost' colour of the spectrum was not red or violet as we now understand them but in fact a series of transitions through 'spectral purple'. On this point he was certainly a little confused. Nevertheless, from this point of view his original statement and the inclusion of purple in his table of complementaries are perfectly consistent. In any case, this work did have the very important result of correcting Helmholtz's own error, which after all, had been the original inspiration for the research. As Grassmann correctly suspected, there are many pairs of complementary colours, which lie opposite each other on a line passing through white.

THE MATHEMATICS OF COLOUR VISION

Grassmann had now established several important points based on empirical observation: (i) the number of variables in colour vision is three, being specifically hue, saturation and brightness. In modern terms these are better denominated as dominant wavelength, purity and luminance; (ii) a continuous alteration of one of two mixed lights gives a continuous variation in colour sensation, and (iii) every colour can be matched by a definite spectral light (or purple) mixed with white light. These statements, now known as Grassmann's Laws, today provide the basis of much research in colorimetry.

While Grassmann had formally stated the first two laws, the third had so far been given only by implication in his photometric thought experiment. He now returned to this point with a formal statement of his third law in somewhat different terms than it is now known:

> Two colours, each of which has a constant hue, and a constant intensity of the intermixed white, also give constant mixed colours, no matter of what homogeneous colours they may be composed. [1, p260]

Once again this seems intuitively obvious; nevertheless this was the first explicit statement of the principle. Grassmann himself felt that the 'proposition . . . appears to be sufficiently proved by the preceding observations.' [1, p260] He did not mean proved in a formal sense for he recognised that ultimately the validity of assumptions is only tested experimentally. Yet he did regard the assumption as well justified by the preceding discussion. However, he had not yet answered the important question concerning the mathematical form appropriate to colour mixtures. Thus the principle was formally stated together with additional argument in order to achieve this important goal. In this sense, what follows is really an additional plausibility argument by Grassmann to justify the principle just stated and to illustrate its important consequences. The result was a new insight into the meaning of colour mixtures of light.

The idea employed was partly implied earlier by Newton's colour circle. A given colour (hue) of a given brightness was to be represented by a line pointing in a given direction and having a certain length. The length of the line could be used to represent one of the other two variables, brightness of the colour or saturation. But Grassmann had already shown that even saturation was to be understood in terms of brightness so he allowed the length of the line to represent brightness of the colour. (This differed from Newton who, in Grassmann's terms, allowed the length of the line to represent degree of saturation.) For example, two luminances of the same colour might have the ratio m. Then one can be represented by a line x units long pointing in a given direction while the second will be represented by a line mx units long in the same direction.

Now suppose we wish to combine two different colours (see Figure 6.4).

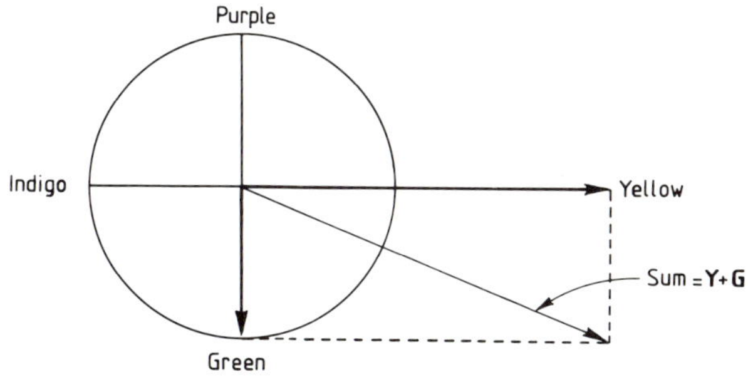

Figure 6.4 Vector addition of two colours according to Grassmann.

Then each one is represented by a directed line of given length originating at the centre of the circle. The compound colour must lie somewhere between the two components, a fact mentioned by Newton and confirmed by Helmholtz provided the colours are close together. It too must be indicated by a directed line of some length. These conditions will be satisfied if the combination is found by determining the diagonal of the parallelogram of which the two colours form the sides. The diagonal points to the resultant colour while its length represents the brightness of the colour. Thus, the very important implication of the principles just stated is that additive combinations of colours of light behave mathematically like vectors. The result of such colour combinations is found by determining their vector sum. This idea was shown in more detail by Grassmann (see Figure 6.5). There the colour marked **c** is composed of the two pure colours yellow and green. The components of **c** are denoted by the broken lines. Similarly the less bright colour, $\mathbf{c}_1$ is a mixture of yellow and green whose components are also shown.

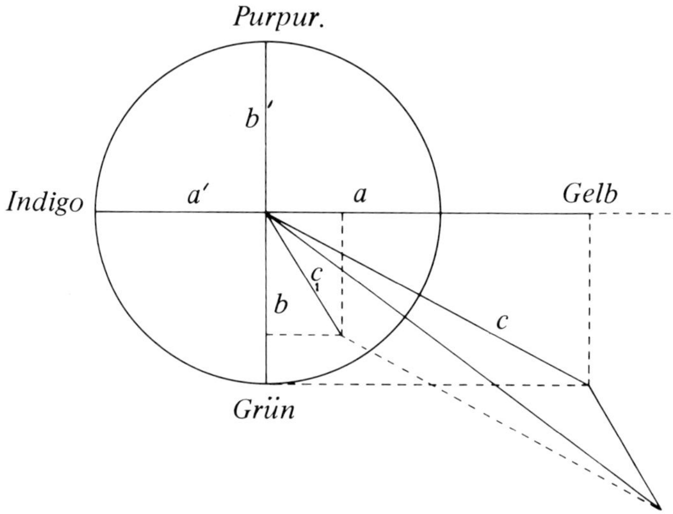

Figure 6.5 Grassmann's vector addition of homogeneous and compound colours. The hues labelled **c** and $\mathbf{c}_1$ are each the vector sum of two homogeneous colours, yellow and green. The long diagonal of the large parallelogram is the vector sum of the compound colours **c** and $\mathbf{c}_1$.

The vector addition of **c** and $\mathbf{c}_1$ is shown in the parallelogram whose long diagonal represents their sum. In analytical terms Grassmann expressed this idea as follows: let there be two levels of brightness of each homogeneous colour; for **a** these will be α and α', and for **b** these will be β and β'. Let one mixed colour, **c**, be represented by

$$\alpha\mathbf{a} + \beta\mathbf{b} = \mathbf{c}. \quad (6.1)$$

Let another mixed colour, $\mathbf{c}_1$, be represented by

$$\alpha'\mathbf{a} + \beta'\mathbf{b} = \mathbf{c}_1. \quad (6.2)$$

Now **a** and **b** are constant monochromatic colours, and **c** and $\mathbf{c}_1$ are constant, but different, mixed colours. A combination of **c** and $\mathbf{c}_1$ must also yield a constant colour. Mixing **c** and $\mathbf{c}_1$ is equivalent to the *vector* addition

$$(\alpha + \alpha')\mathbf{a} + (\beta + \beta')\mathbf{b} = \mathbf{c} + \mathbf{c}_1 = \mathbf{D}. \qquad (6.3)$$

The colour **D** is of equivalent hue and luminance as required since it is composed of the four components, $\alpha\mathbf{a}$, $\alpha'\mathbf{a}$, $\beta\mathbf{b}$, and $\beta'\mathbf{b}$. The total luminance of colour **a** is $(\alpha + \alpha')$ and that of **b** is $(\beta + \beta')$. One simply adds the two rectangular components of each colour to find the total luminance of that colour. Referring to Figure 6.5 the lengths of each component of the long diagonal are each equal to the sum of the two shorter sections of which they are composed. Note that in this addition, the colours **a** and **b** play the role of unit vectors. The very important meaning of equation (6.3) is that a mixture of any two different colours, **a** and **b**, can be made equivalent in appearance to a third colour, **D**. It provides an operational basis for photometrically matching different colours and the subsequent specification of those colours in terms of vector equations.

Grassmann's statement today is understood in another form. If two lights of the same appearance are added to two other (but different) lights of *equal* appearance, the resulting mixed lights will still appear to be matched although the particular colour impression will be different. Put otherwise, suppose we have two originally matching (equal) coloured lights, **a** and **b**. Let there be another set of matching lights, **c** and **d**. Then a mixture of $\mathbf{a} + \mathbf{c}$ will have the same appearance (or look like) a mixture of $\mathbf{b} + \mathbf{d}$. This is an application to colour mixtures of light of the axiom stating that if equals are added to equals the results are equal. [10] It follows also that subtraction of equal colour components from two matching colours must also leave them still equivalent in appearance although again with a new colour sensation. And increasing or decreasing the brightness of equivalent colours (multiplication and division of equals) also leaves the apparent colour equivalence unchanged provided they are not made too bright or too dark.

Having now illustrated the vector nature of compound colours (of light) Grassmann went on to show that this understanding and Newton's centre of gravity method of combining colours were identical. And here a more subtle motivation for Grassmann's work appears than its ostensible scientific goals. A Gymnasium (high school) teacher in Stettin, Germany, Grassmann had published in 1844 a work entitled *Die Lineale Ausdehnungslehre*. Together with a similar work by William Rowan Hamilton (published after Grassmann's) it ultimately led to the creation of vector and tensor analysis. (Grassmann's school little knew the talent it harboured.) Indeed it was Grassmann's work more than Hamilton's that led to modern vector analysis. But that lay many years in the future; nine years after its publication, Grassmann's book still lay sunk in obscurity. To his deep disappointment his profound effort, later described as one of the supreme landmarks of the human mind, [11] fell absolutely flat on the scientific community.

Almost certainly Grassmann's main problem in achieving scientific recognition was a difficult compressed writing style. Added to this was much abstruse, distracting discussion in his *Ausdehnungslehre* on the philosophy of mathematics making the work truly forbidding. Even so able a mathematician as A F Möbius, Grassmann's friend, excused himself for only a slight perusal of the work saying that he could not follow Grassmann's thought, and Hamilton who more than anyone else could appreciate Grassmann's genius said that the book was so difficult it had driven him to smoke! It required a much revised second edition in 1862 and yet another generation for Grassmann's ideas to filter into the mainstream of mathematics. Hoping all his life for a university appointment, Grassmann remained a Gymnasium teacher mostly outside the main scientific stream. It seems therefore that Grassmann saw in this work an application of his new vector methods to a problem that anyone could appreciate. And he also hoped to alert the scientific community to his neglected *Ausdehnungslehre*. Quoting a paragraph from his book, Grassmann now applied his new vector methods for finding the centre of gravity of any number of mass points to Newton's colour circle.

Recall that Newton had assigned a mean colour to each colour sector whose position was marked by a tiny circle. The brightness of each colour was represented by the area of the circle, which in turn was treated as a loaded point whose weight represented the brightness. Then the object was to find the centre of gravity (more properly the centroid) of the little circles.

As we have seen Grassmann drew a circle of colours like Newton's but with its divisions altered. A line from the centre of Grassmann's circle represented a given homogeneous colour by its direction, while its length represented the brightness. Instead of mean colours, Grassmann used any particular colour involved. For each line he substituted a weight on the circumference at the place indicated by the line's direction. The amount of weight was proportional to the line's length. As Grassmann said:

> Each of the colours is represented by a loaded point on the periphery, in such a manner that the radius belonging to it shows its hue, whilst the weight expresses the intensity, and determines the centre of gravity. [1, p262]

Then the hue of the compound colour was determined by a line from the centre of the circle to the centre of gravity of the homogeneous components, and the brightness of the compound colour was determined by multiplying this length of line by the sum of the weights.

This procedure is equivalent to the preceding vector addition. Referring to Figure 6.6 we have any number of mass points, A, B, . . . of masses m_1, m_2 . . . One draws the radius vectors OA, OB . . . which are denoted respectively by $\mathbf{r}_1$, $\mathbf{r}_2$ . . . The radius vector of the centre of gravity of these points is denoted by $\mathbf{R}$. Then according to Grassmann, the centre of gravity with respect to O is given by

$$\mathbf{R} = \frac{m_1\mathbf{r}_1 + m_2\mathbf{r}_2}{m_1 + m_2} . \tag{6.4}$$

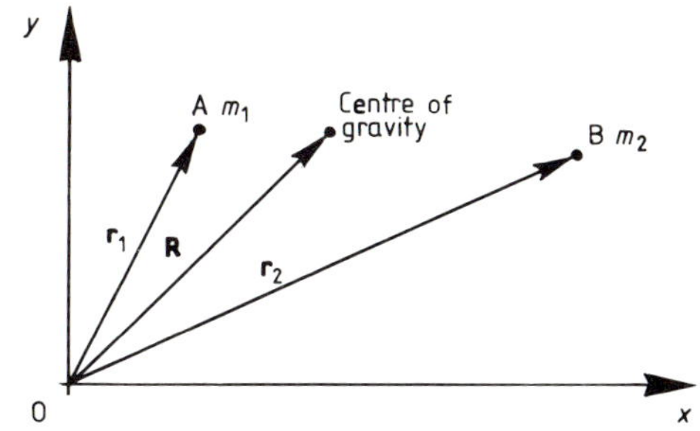

Figure 6.6 Radius vectors to locate the centre of gravity between two mass points.

But each radius vector can be put in rectangular coordinates in terms of the unit vectors **i** and **j**. Thus

$$\begin{aligned}\mathbf{r}_1 &= x_1\mathbf{i} + y_1\mathbf{j}\\ \mathbf{r}_2 &= x_2\mathbf{i} + y_2\mathbf{j},\end{aligned} \tag{6.5}$$

and we can write equation (6.4) as

$$\mathbf{R} = \frac{m_1(x_1\mathbf{i} + y_1\mathbf{j}) + m_2(x_2\mathbf{i} + y_2\mathbf{j}) + \ldots}{m_1 + m_2 + \ldots}. \tag{6.6}$$

Restricting ourselves to two points this can be written as

$$(m_1x_1 + m_2x_2)\mathbf{i} + (m_1y_1 + m_2y_2)\mathbf{j} = (m_1 + m_2)\mathbf{R}. \tag{6.7}$$

To obtain the coordinates of the centre of gravity one need only express **R** in terms of its components and equate the terms in **i** and **j**. Now compare this result with equation (6.3). We can identify the colours, **a** and **b**, in that equation with the unit vectors **i** and **j** in equation (6.7). The brightness coefficients multiplying each colour unit vector, **a** and **b**, are analogous to the mass coefficients multiplying the rectangular unit vectors above; and the compound colour **D** is analogous to **R** in equation (6.7) where the sum of the masses multiplying **R** is equivalent to the sum of the colour brightnessses of the compound colour. Thus Grassmann had achieved his ultimate goal, showing that the new concept of colours as vectors was formally identifiable with Newton's centre of gravity method of determining mixed colours. Therefore Newton's original method, now revised, was a suitable basis for a mathematical theory of colour mixtures. For Grassmann these conclusions followed of necessity from his original assumptions which, as stated before, he felt were already well justified. Still, he recognised that assumptions need to be tested, so he concluded this argument by saying that the only requirement yet outstanding was 'a simple but complete series of observations for the determination of the series of colours.' [1, p262] This was accomplished partly by Helmholtz less than a year later but mostly by Maxwell in 1855 and 1860.

Although Grassmann had accomplished his major objectives at this point there was still a logical omission to the theory. His modification of Newton's system with vector methods accounted for only two of his three colour variables, hue and luminance. So far he had not shown how his 'intensity of intermixed white' could be worked into his description. This inconsistency he accounted for with a fourth and final assumption that:

> the total intensity of the mixture is the sum of the intensities of the lights mixed. [1, p263]

Now Grassmann wanted simple addition of the light intensities. But light intensity is proportional to the square of wave amplitude while the simple addition required here implied linear dependence. Recognising this difficulty Grassmann specifically qualified his term 'intensity' to mean linear proportionality, not proportionality to the square of amplitude as usually understood. [1, p263] However, Grassmann's term 'intensity' must mean in modern terms 'luminance' which is itself intensity per unit area and is therefore proportional to wave amplitude squared. Moreover, the term intensity is strictly appropriate only for point sources whereas Grassmann was concerned here with extended areas of colour. And so Grassmann's nomenclature was confusing, which difficulty undoubtedly arose from rather imprecise terminology in photometry at the time. Although photometry was already a well known science it had yet to clarify some of its basic terms. Given Grassmann's understanding, however, his term 'total intensity' then included the sum of the luminances of each colour plus that of the intermixed white. [1, p264] However, considering Grassmann's previous assumptions, this hypothesis is essentially a tautology; for it is equivalent to the conclusion drawn from equation (6.3) which adds the luminances of the colours. Indeed it is really a tacit assumption of that vector addition differing from that conclusion only in specifically including the intermixed white (saturation) of the colour. Nevertheless, Grassmann thought this assumption was not so well founded in observation unlike his previous statements. Still it seemed to be not only the simplest supposition possible but also the most probable and subsequent events have borne this out. Luminances of different coloured lights including white are simply additive in Grassmann's sense, an observational fact underlying the very science of colorimetry. From this final hypothesis, then, Grassmann accounted for the saturation of the compound colour.

The circle of Figure 6.7 represents the usual circle of colours with centre O. For simplicity Grassmann supposed the radius of the circle to represent unit intensity. Points A and B on the circumference represent two homogeneous colours. When on the circumference they represent also unit intensity. If their intensities are taken actually as α and β respectively, then we have two colours, αA and βB to compound. Then each point is considered to have a weight attached to it and we find the centre of gravity by the usual process. Suppose C is the centre of gravity, then at C the intensity of the compound colour is $(\alpha + \beta)$OC. But the total intensity of the compound hue *and* intermixed white is written in terms of the radius OD. According to the

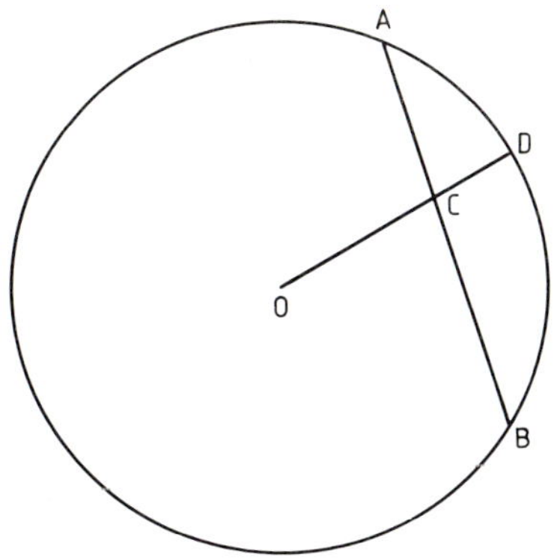

Figure 6.7 Grassmann's diagram to account for colour saturation.

hypothesis, that total intensity (luminance) is $(\alpha + \beta)$OD, and we can write

$$(\alpha + \beta)\mathrm{OD} = (\alpha + \beta)\mathrm{OC} + (\alpha + \beta)\mathrm{CD}. \qquad (6.8)$$

The term on the left had previously been identified as the total luminance (colours plus intermixed white) while the first term on the right was identified as the intensity (luminance) of the compound colour. The remaining term, according to the hypothesis, must be the intensity of the intermixed white, or the brightness of the saturation. Grassmann concluded that 'the intensity of the intermixed white is . . . equal to the distance of the centre of gravity from the periphery, multipled by the sum of the weights.' [1, p264] Not unexpectedly this conclusion conforms with Newton's understanding of saturation in his circle. There maximum saturation lay on the circumference and maximum desaturation (white) lay at the centre. Here as the colours A and B are brought closer and closer together, the point C approaches the circle and the length CD approaches zero in the limit. The luminance of the white is therefore essentially zero near the circumference. Conversely as A and B are taken farther apart, point C approaches the centre of the circle, which from the argument on complementary colours represents pure white. When they are opposite each other along a diameter, A and B are complementary, CD is maximum and the desaturation is maximum. This conclusion is implied, at least, in the previous proof of complementary colours so the 'proof' here and the assumption on which it stands can be considered as unnecessary. Still the idea of saturation had not been *explicitly* incorporated into the general scheme thus far and Grassmann clearly felt that this plausibility argument was preferable to implication.

It is interesting to compare Grassmann's concepts with the colour pyramids of Mayer and Lambert. The base of the pyramid was a triangle of three primary colours (not to be confused with the three variables defined by Grassmann). Combinations of the primary colours lay within the triangle by finding centres of gravity in the sense of Newton. These mixtures Lambert had noted were paler or less saturated than the primaries so the centre of the triangle tended towards grey. Then both Mayer and Lambert added white to each primary and formed a new colour triangle of shorter sides than the previous one. The second triangle was placed on top of the first and new colour combinations could be found as previously except that they lay above the base triangle. Repeating the procedure several times one reached the apex of the pyramid which was entirely white. Lambert had also conceived a second pyramid darkened toward black at its apex. Placed base to base with the other, continuous transitions in brightness would be possible from the bright white apex of one pyramid to darkness at the black apex of the other.

From Grassmann's point of view the base of these pyramids represented two variables, hue and brightness of intermixed white, with the luminance of the colours themselves held constant. To alter the luminance of the colours the third dimension of space was required. Adding white to each primary colour in building the pyramid was then tantamount to changing the brightness of the colours. With the primaries diluted with white going toward the apex the ability to discriminate one hue from another was diminished and the sides of each new triangle grew shorter as fewer colour combinations were possible. Eventually the ability to distinguish any colour differences at all was destroyed by the dazzlingly bright light of the apex. (Similarly, darkening the colours in Lambert's second pyramid also destroyed the ability to discriminate colours.) If the pyramid were a cone then a projection of the cone from above the apex onto a plane would give Newton's colour circle, with mixed colours inside the circle and white at the centre. We can see now that Newton's rule for compounding mixed colour employed two of Grassmann's variables, colour and saturation, with colour luminance held constant (see Figure 6.8). [12] The pyramids of Mayer and Lambert then represented a

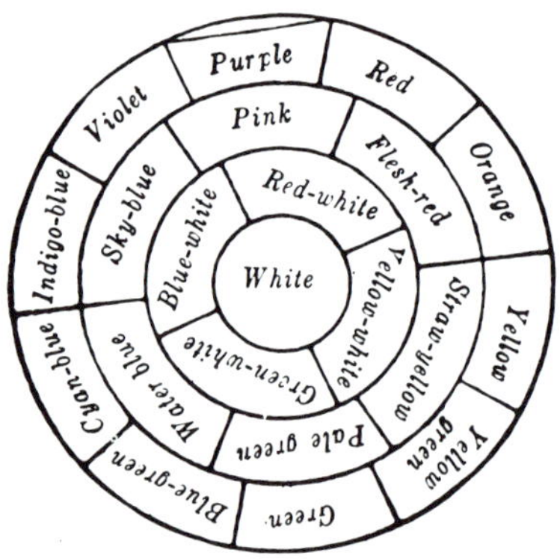

Figure 6.8 Projection of the colour cone from the apex onto the base. The largest circle is then Newton's colour circle. The next two circles represent sections through the cone parallel to the base. The inner circle is a section of the cone from near the apex (see Figure 9.10).

three-dimensional arrangement of colour in terms of Grassmann's three variables but without clearly identifying them. Similarly Runge's sphere and Chevreul's hemisphere each found three-dimensional space both necessary and sufficient to categorise all colour mixtures. In all these works there was a tacit analogy of three-dimensional space with the requirements of colour perception. But there is nothing in three-dimensional space which *ipso facto* points towards a trichromatic visual response mechanism. It is an analogy only and need not follow with compelling necessity. It did, however, imply that only three different modes of modifying colour sensations were known. The analogy was unfortunately lost upon these investigators who uniformly thought in terms of primary colours and not optical and photometric observables. (Mayer and Lambert, of course, could not have thought in terms of wavelengths but ironically Lambert wrote an important work on photometry.) So the trichromatic concept of colour vision remained, until Grassmann, at best only a poorly founded hypothesis. Oddly enough, however, the validity of this space analogy was given credibility by yet another analogy. Grassmann in arguing for three colour variables showed that language itself admits of only three colour distinctions, hue, brightness and saturation. So if words enter a language to express generally recognised concepts, this linguistic situation constituted unrecognised evidence that three variables of colour and no more were already intuitively recognised. Significantly, Grassmann was also a philologist whose important contributions, unlike his mathematics, were recognised during his lifetime.

In this work Grassmann played a rather conservative role *vis à vis* that of Helmholtz. Whereas Helmholtz was willing to give up Newton's colour circle, Grassmann felt that to do so left physics with no viable starting point for constructing a theory of colour. For him it was necessary to preserve Newton's colour circle at all costs. Moreover, Grassmann did not mention the work of any of his predecessors in this area except Newton and Helmholtz. His approach to colour perception was remarkably different from all of them. While they sought simply to summarise the facts of colour combinations (as they perceived them) he sought for optical observables. While their summaries tended to be sterile in terms of physical concepts, his new ideas were ultimately most fruitful. By showing that colours could be viewed as vectors, that they could be measured photometrically and by indicating the possibility of colour equations (in a form much different from Forbes') he laid the foundations for a science of quantitative colour measurement. From this emerged a new physical understanding of colour.

HELMHOLTZ AND THE COMPLEMENTARY COLOURS

If Grassmann's work was stimulated by that of Helmholtz, Helmholtz was in turn prompted by Grassmann. Helmholtz returned to the problem of spectral complementary colours shortly after Grassmann's paper appeared. [13] He acknowledged his error of a single complementary pair ascribing it to the limited field of the V-slit arrangement and physiological problems in the eye caused by it. [9, p1] As luck would have it, however, a much superior method of colour mixing was published by Léon Foucault at almost the same time as

Grassmann's paper. [14] In confirming Helmholtz's first results, Foucault said he had known of this method since 1849, for he had seen it then in physics lectures given by Claude Servais Pouillet at the Sorbonne. It is startling to find, considering the conceptual and experimental difficulties besetting mixed spectral colours, that a good method of combining them had already existed before Helmholtz took up the problem. Almost certainly Pouillet realised some of Helmholtz's 1852 results in 1849 during demonstrations! Foucault in using Pouillet's method, however, did nothing but verify Helmholtz's V-slit results publicly without extending them so the original question concerning complementary colours remained. But in publishing the method itself he performed an important service since it was this that allowed Helmholtz to return to the question so rapidly.

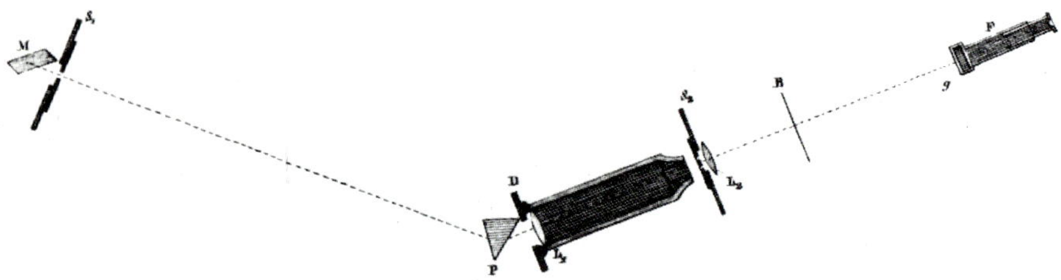

Figure 6.9 Helmholtz's experimental arrangement of 1853 to determine spectral complementary colours together with their wavelengths.

Helmholtz employed the new method as shown in Figure 6.9. [9, Table 1, Figure 2] Sunlight reflects from a heliostat, M, and passes through slit S_1 to the prism, P, about 10 ft away. The resulting spectrum then passes through a diaphragm D and achromatic lens, L_1, which focuses the spectral image on a screen, S_2. This screen, made of brass, carries two mechanically adjustable slits which enable the observer to isolate any two separate portions of the spectrum. Then the achromatic lens, L_2, focuses the selected colours onto one area of the white card at B. This provides a much larger field than the earlier experiments while providing adjustable control over the colours observed. The V-slit experiments, by comparison, fixed the selection of colours by the intersection of the slits. The telescope at F was used in a second part of the experiment to measure the wavelengths of the complementary colours by comparison with known Fraunhofer lines. With this arrangement Helmholtz determined seven complementary pairs of spectral colours together with their wavelengths. These are shown in Table 6.1. [9, p15] [15]

The distribution of the complementary wavelengths in the spectrum was

Table 6.1. The seven pairs of complementary spectral colours determined by Helmholtz.

Colour	Wavelength (nm)	Complementary colour	Wavelength (nm)
Red	656.2	Green-blue	492.1
Orange	607.7	Blue	489.7
Golden Yellow	585.3	Blue	485.4
Golden Yellow	573.9	Blue	482.1
Yellow	567.1	Indigo-blue	464.5
Yellow	564.4	Indigo-blue	461.8
Green-yellow	563.6	Violet	from 433 on

most unexpected, however. Helmholtz plotted the wavelengths as shown in Figure 6.10, forming a curve of two branches. [9, Table 1, Figure 3] [16] There as one proceeds along the abscissa from violet to red the change in complementary wavelengths is rather gradual but on reaching the greenish-blue region there is a rapid variation of wavelengths and therefore of apparent hue. Similarly from red towards violet a gradual change of complementary wavelengths takes place until the yellow where the changes are also rapid. As expected, there are no points in the greens indicating that no complementary colours were observed for these colours. The apparent hue at the ends of the spectrum changes rather slowly as compared with wavelength but it changes very rapidly in the middle of the spectrum. While there is a fairly large space between red and yellow, their complements, green-blue to cyan-blue, are located very close together. Similarly, violet and indigo occupy a relatively broad portion of the spectrum while their

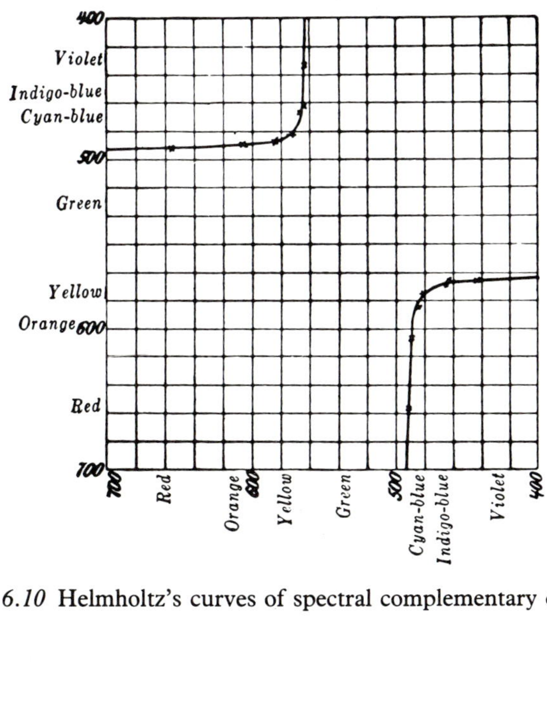

Figure 6.10 Helmholtz's curves of spectral complementary colours.

complements, in the yellows, occupy narrow strips. Considering this it is clear that in a small-area field with crowded lines it would have been difficult indeed to recognise complementary pairs that changed appearance so rapidly. Moreover, widely separated complements such as orange and blue, having such different refrangibilities, are not easily focused together by the eye. Small wonder that Newton with limited means was unable to obtain white from only two spectral colours.

In addition to these results, Helmholtz also estimated the relative luminances of the complementary colours pointing out that they need not be equally bright. Equal brightness of the two colours was, of course, assumed by both Newton and Grassmann. Helmholtz too had insisted on this with his yellow and indigo-blue pair in 1852. But these results were obtained in several instances with unequal slit widths at S_2 and therefore unequal luminances judged by the eye. This was not unexpected considering his earlier observations with slips of paper in front of the V-slits to change brightness. But he had thought then that such a procedure constituted an artificial change in brightness and so he had rejected the observations. Hence he had settled on his one 'natural pair'. Now, however, by measuring slit width, he estimated relative brightnesses and rejected his earlier assumption requiring equal luminance. The violet and green-yellow pair, for example, had a brightness ratio of 1:10 respectively in bright light while the ratio of indigo to yellow was 1:4 respectively. [9, p19] Therefore indigo and violet appeared darker than their complements which implied that they were more saturated than their complements. From similar estimates on the other pairs, Helmholtz concluded that the relative order of saturation in the spectrum would be as below with violet the most saturated: [9, p21]

Violet
Indigo-blue
Red Cyan-blue
Orange Green
Yellow

Such results were unsuspected, of course, by Grassmann whose original prediction concerning complementary colours however was confirmed. But Helmholtz now said that he must be wrong about the circular locus of the colours. There the colours were all equally far from the centre which represented white because they were tacitly assumed to be equally saturated; hence equal brightnesses could be used. Based on this Grassmann had shown that colours at opposite ends of a diameter through white must be complementary. If this idea were to be preserved, the more saturated colour had to lie further from the white point than its complement. Violet, for instance, had to lie about 10 times farther away from white (for bright light) than green-yellow. Finally, allowing for purple transitions along a straight line, which Grassmann omitted, Helmholtz estimated the curve of spectral colours to be like that in Figure 6.11. [9, p19] [17] Considering the gradual change of the hues at the ends of the spectrum, Helmholtz assigned those parts of the

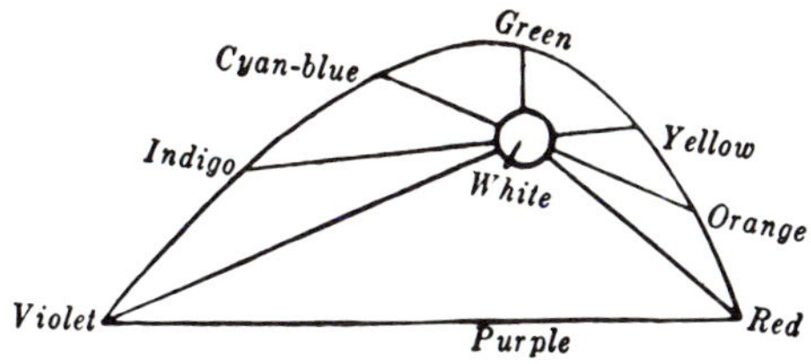

Figure 6.11 Helmholtz's hypothetical curve of the locus of spectral colours.

curve a relatively low curvature, and since the hues change rapidly in the yellow–green-blue section he assigned this section a larger curvature. Figure 6.11 shows that a wide range of greens have no natural complements, their complementary colours lying along the purple chord. How this reasoning compares with more modern ideas will be seen in Chapter 9. For now Helmholtz's proposed curve served, in effect, a heuristic function but did not represent an actual determination of its shape. It still remained to derive the locus of the spectral colours experimentally, which was only done by James Clerk Maxwell in 1860.

NOTES

1 Grassmann, Hermann Günther 1853 Zur Theorie der Farbenmischung *Ann. Phys., Lpz.* **89** (English translation 1854 *Phil. Mag.* **7**). Reference to Grassmann's paper are to the English version. A much abridged and somewhat revised version of the English translation appears in MacAdam, David L (ed) 1970 *Sources of Colour Science* (Cambridge, Ma: MIT Press)

2 The term in the translation is 'intensity', which is better understood as 'brightness' or in modern terms as 'luminance'

3 Abney, William deW 1910 On the change of hue of spectrum colours by dilution with white light *Proc. R. Soc.* A **83** 120

4 Parsons, J Herbert 1915 *An Introduction to the Study of Colour Vision* (London: Cambridge University Press) p33

5 Newton, Isaac 1730 *Opticks* 4th edn (London). Reprinted 1952 (New York: Dover) p156

6 Grassmann admitted that the assumption that Newton's boundaries coincided with Fraunhofer's B and H lines was somewhat arbitrary. However, he pointed out that the same result could be obtained by assuming the mean refractive indices of Newton's and Fraunhofer's prisms to coincide for each different colour. These were the only two possibilities for his programme of revising Newton's colour circle as a basis for a mathematical theory of colour

7 von Helmholtz, Hermann 1852 On the theory of compound colours *Ann. Phys., Lpz.* **9** (English translation 1852 *Phil. Mag.* **4**)

8 Grassmann incorrectly cites *Ann. Phys., Lpz.* **13** 441 but this should be *Ann. Phys., Lpz.* **23** 441. This reference was to the French chemist Jean-Henri Hassenfratz, whose own observations dated back to 1799. See further Hassenfratz, J-H 1808 Mémoire sur les alterations que la lumière du soleil éprouve en traversant l'atmosphere. *Ann. Chem.* **66** 54–62

9 von Helmholtz, Hermann 1855 Uber die Zusammensetzung von Spectralfarben *Ann. Phys., Lpz.* **94**

10 See, for example, Judd, Dean B and Wyszecki, Gunther 1963 *Color in Business, Science and Industry* 2nd edn (New York: Wiley) pp48–9

11 Sarton, George 1944 Grassmann *Isis* **35** 326. Sarton gives a short account of Grassmann's career in comparison with that of William Rowan Hamilton. For a more extended and excellent account of Grassmann's career and his problems in gaining recognition by the scientific community see Crowe, M J 1967 *A History of Vector Analysis* (Southbend, Indiana: University of Notre Dame Press)

12 von Helmholtz, Hermann 1911 *Physiological Optics* 3rd edn (Hamburg and Leipzig). English translation Southall, James P (ed) 1924 Vol II (New York: The Optical Society of America) p132

13 The experiments reported in [9] were mostly performed between later May and September, 1853. Grassmann's paper was submitted to *Ann. Phys., Lpz.* by 7 May 1853 and Helmholtz gave a preliminary report of his results at the meeting of the British Association held in Hull in September 1853. See *Rep. Br. Assoc.* 1853, Transactions of the Sections p5

14 Foucault, Léon 1853 Sur la recomposition des couleurs du spectre en teintes plates *Cosmos* **II** 232

15 Helmholtz originally measured his wavelengths, as was the custom then, in Fraunhofer's units, millionths of a Paris inch. These values were republished in 1860 and again in the first edition of *Physiological Optics.* [12] In the second edition they were converted to millionths of a millimetre (equivalent to nanometers) and it is these values that are used as being more meaningful now. J von Kries extended Helmholtz's results to 13 pairs in 1881 and similar results have been obtained by others as well

16 The curves shown here are not those from the original paper. [9] As noted above [15] the units were later converted to nanometers and so the curves were redrawn in later publications. Figure 6.10 is from the English translation of *Physiological Optics*. These curves are essentially the same as Helmholtz's originals of 1853

17 For consistency the English version of *Physiological Optics* has been used. The figures are the same in all respects

7
Interlude: The Discovery of Colour Blindness

Intelligent communication in language is based on general agreement on the meaning of words. In terms of sensations, agreement on words for taste, touch, sound and vision implies that everyone experiences essentially similar impressions under similar conditions. Were this not so Grassmann's linguistic arguments for three colour variables would be necessarily false. Yet in respect to vision in colour this is in fact not always so. Many individuals exist for whom colour sensations are distinctly abnormal and for whom certain obviously different colours appear identical. Red and green, for example, may not be discriminated from each other but may actually look exactly alike. Hence two objects separately coloured red and green will appear equivalent and be indiscernible barring other visual clues. A red cloth placed on green grass blends into the background and is invisible to such persons. This visual defect, commonly called colour blindness, is much more subtle than ordinary optical defects of the eye such as myopia or hypermyopia. Those difficulties, being related to object–image distances, at least give the individual some point where vision is clear and sharp. The affected person has one standard by which to compare his vision under varying distances. From this simple fact evolved the lens grinding trade and the origins of spectacles as early as the fourteenth century. But colour blindness is generally difficult for the afflicted person to detect within himself as his visual acuity in other respects may be perfect. His only clue to skewed colour perception will be occasional disagreement in conversations involving colour and he can reasonably think that other people are at fault rather than himself. Lacking any objective standard of comparison, the colour blind person can easily go through life never suspecting his own difficulties. Nevertheless, he must in fact remain ignorant of colours as compared with normal sighted persons. How much more difficult it is then for normal persons to detect colour blindness in others. In essence communications on colours between normal and colour blind individuals becomes impossible. For this reason colour blindness as a deviation from normal vision was hardly recognised before the unique self-analysis of John Dalton in 1794.

A survey of ancient and medieval writers on optics and vision shows that they did not discuss vision in terms of colours at all and therefore colour blindness was unrecognised by them. [1] The greatest optical writers in the seventeenth century, Kepler, Gassendi, Huygens and Newton, also made no reference to such anomalous vision. An oblique reference to possible colour perception problems was made by Philippe de la Hire in 1678 caused by tinting of the humours of the eye but he was thinking in terms of injury and cited no examples of any sort. [2]

Robert Boyle, however, did note in passing a case of genuine colour blindness in 1688. Describing a conversation with an eminent, but unnamed, astronomer Boyle remarked,

> he confesses to me, that there are some colours, that he constantly sees amiss, and particularly instanced in one, which in a clear day (for so it was when we discoursed together of this matter) seemed to him to be the same with that of a darkish sort of cloth, that he then wore, whilst to me and other men it appeared of quite differing colour. [3]

What these colours were Boyle did not say but the fact that there was an apparent identity of two distinct colours marks this person as being genuinely colour blind. However, Boyle simply reported the remark without comment as only one of fourteen cases of various eye problems many of which were cataracts. Such indiscriminate lumping together of diverse optical difficulties by one of the foremost men of seventeenth-century science indicates that most visual problems were not well discriminated from each other (except myopia and hypermyopia), and cataracts, of course, had been known since Antiquity. Clearly colour blindness went undiscovered in the sevententh century, Boyle notwithstanding.

The eighteenth century was hardly different, for developments in optics and physiological optics during that century (and earlier) were described in detail by Priestley in 1772 with no reference to people who saw colours abnormally. [4] Nor did Priestley even mention Boyle's observation. Therefore, it is clear that colour blindness was still unrecognised as late as 1772. Indeed, this conclusion is confirmed by a survey of the leading journals of these times where the subject in any form is conspicuous by its absence. In the *Philosophical Transactions*, for example, (an excellent guide to topics under general discussion) there were from its inception in 1665 to 1800 only three short articles about persons who saw colours improperly. Moreover, the first of these, appearing in 1684, is a dubious case, probably better classified otherwise as can be seen from the following description:

> A Maid, two or three and twenty years old, came to me from Banbury, who could see very well but no colour beside Black and White. She had such Scintillations by night, (with the appearances of Bulls, Bears, etc.) as terrified her very much; she could see to read sometimes in the greatest darkness for almost a quarter of an hour. [5]

Significantly, the next article on colour vision in the *Philosophical Transac-*

tions was in 1777, [6] five years after Priestley's survey and 93 years after the above description. Here a man identified only as Harris described his difficulties to Joseph Huddart the author of the article who was a well known hydrographer and manufacturer. Most interestingly, considering the necessity sometimes for linguistic evidence, Harris, like Grassmann, was alert to distinctions in language. He noted an apparent difference in the vocabulary of colours between himself and other persons. Huddart wrote that Harris

> had reason to believe other persons saw something in objects which he could not see; that their language seemed to mark qualities with confidence and precision, which he could only guess at with hesitation, and frequently with error. [6, p261]

Indeed, Harris was a rather good introspective observer and Huddart a good recorder of some difficulties of colour blind people. For example, Harris stated that he experienced difficulties, compared to others, in seeing cherries on a tree. While his companions could apparently distinguish the cherries at a distance by their colour he could not distinguish between the leaves of the tree and the cherries except by differences in size and shape, and for this, of course, he had to be relatively close. In situations less dependent on colour, however, Harris observed that his visual acuity was as sharp as anyone else's. Therefore Harris gave the first intimation that visual discrimination between objects depends upon colour differences as well as size and shape. A person seeing different colours as equal lacks an important visual clue. Since this visual problem was so extraordinary Huddart noted carefully that Harris was an otherwise intelligent man implying thereby that his colour perception problems stemmed from the eye itself and not, as some might think, from inferior mental capacity.

Harris' brother, master of a sailing vessel, and also presumably not retarded, was also colour defective and allowed Huddart to test his perception with bits of coloured ribbon viewed in bright daylight. Asked to identify the colours of the ribbons, he hesitantly identified light green as yellow and pale red as blue. But of an orange stripe he confidently said: 'this is the colour of grass; this is green.' [6, p267] For other colours he seems to have been normal. The point here is that these cases described by Huddart constitute the first published account of genuine cases of colour blindness. Remarkably, Huddart did not seem to recognise the real discovery he had made. The general tenor of his article is one of justifiable curiosity together with amazement. He could hardly believe what he was seeing before his eyes. Understandably he could doubt at first the intelligence of his subjects. So in the end he failed to realise that his cases were not singular events, unlike so many of the odd curiosities reported in the *Philosophical Transactions* during this period.

These case histories were picked up by l'Abbé Rozier in 1779 in his journal *Observations sur la Physique, sur l'Histoire Naturelle et sur les Arts*. Describing the Harris brothers, Rozier added another case of the minor French poet Colardeau who could not distinguish colours. Although greatly skilled in sketching, Colardeau required someone else to select his colours when he wished to paint. Left to himself he had used bright red which he thought was deep blue, he used yellow in place of blue, and placed red next to green without seeing the difference. Noting now three similar cases of colour defective vision, Rozier pointed out that perhaps this kind of vision was not so rare as one might think; as to its cause, however, he did not speculate.

Huddart's article made an impression on at least one reader of the *Philosophical Transactions* who elicited another case history in 1778 as a result. Of this person, one J Scott, one significant paragraph characterised the problem. Describing himself to the author, Scott said:

> I do not know any green in the world. A pink colour and a pale blue are alike, I do not know one from the other. A full red and a full green are the same, I have often thought them a good match; but yellows (light, dark and middle) and all degrees of blue, except those very pale, commonly called sky, I know perfectly well . . . A full purple and deep blue sometimes baffle me. [7]

Scott was apparently defective in red vision having limited visual sensation at the red end of the spectrum. Objects in that range of colours would appear black to him, a point he unwittingly emphasised with an anecdote. Relating how his daughter's fiancé dressed himself in black the day before the wedding, he was much distressed at the young man's poor taste until his daughter advised him that the colour was in fact claret.

In addition to his own symptoms, Scott most interestingly described three generations of his family who had either encountered the same difficulty or escaped it so that from his genealogy there was a strong implication that the defect was an inherited one. Both his father and maternal uncle had the defect although his mother did not. However, of his two sisters one was also colour defective so that Scott recorded a rare case of female colour blindness, resulting in modern terms, from a female carrier and a colour blind male. As might now be expected both the sons of this sister had the defect while his own son and daughter did not. Scott was then considering the descent of the problem to his grandchildren, a fourth generation beginning with his parents, but sadly as yet 'had none living' although his daughter was then married for several years.

The cases of Harris and Colardeau were mentioned again in 1781 in J C Lichtenburg's magazine, which in turn took the information from Rozier. Beyond mentioning the symptoms, the article also referred to a severe case of 'an apothecary in Strassburg' without otherwise describing his difficulties. [8] Nevertheless, Huddart's article had now been noticed in France and in Germany. Huddart's article was noted again with extensive quotation in the 3rd edition of the *Encyclopaedia Britannica* in 1792 while Scott's symptoms, too, were briefly recorded under a separate heading. [9] Both were simply stated without comment although the same quiet amazement appears in the introduction to Huddart's article as in Huddart himself. The problem was alluded to once more in 1792 by the Scottish philosopher Dougald Stewart who remarked, in passing, that differences in colour perception existed in men. He ascribed it simply to poor training in colours but strangely enough he was himself colour blind not being able to distinguish between the green leaves of an apple tree and the red apples. [10] (One wonders whether he ever tried to improve his own 'training' in colours.) None of these articles, however, resulted in any new information for they mostly relied on Huddart's original description while only mentioning superficially a new case of their own. Unlike Huddart, the new cases mentioned in Rozier's and Lichtenburg's journals were not personally investigated by their authors. So as of 1792, 15 years after its original publication, Huddart's article was still the chief source of information on the subject. Still it is significant that the problem was worthy of mention at all. The continual references indicate at least some awareness of a new and unusual visual problem; and by adding new cases, even though poorly described, they began to accumulate some statistical evidence. But it is also clear from these short notices that no real appreciation of colour blindness existed and that research in abnormal colour perception hardly existed as a scientific concern as late as 1792.

It was John Dalton, the eminent chemist, who gave the first detailed description of colour blindness in 1794 concerning his own problem. [11] It is certainly a fortuitous accident that a man of Dalton's scientific perception and integrity happened to be colour blind. Without this stimulus, the general problem would have remained unrecognised even longer; and without Dalton's later scientific reputation his paper would have been ignored completely. We shall see presently that the general problem was largely ignored anyway by most eminent scientists of the early nineteenth century, only a few deeming it worthy of research.

Before undertaking his specific investigations, Dalton had received intimations of his problem. He had given his mother a pair of apparently dark blue stockings in which to attend Quaker meetings. To his amazement and dismay she told him they were 'red as a cherry' although he frankly thought the discrepancy lay in her aged eyesight and not his own. However, her neighbours all concurred that they were 'uncommon scarlety' and Dalton was forced to acknowledge something amiss in himself. [12] Similarly, writing to his cousin in February 1794 Dalton tells how he amused his friends by calling his green coat red and confusing several other colours. By now, however, he was explicitly aware of his problem which had been finally brought home to him in 1792 when he accidently observed a pink geranium under candle light. Describing that encounter Dalton said,

> I was never convinced of a peculiarity in my vision, till I accidently observed the colour of the flower of the Geranium zonale by candle-light, in the Autumn of 1792. The flower was pink, but it appeared to me almost an exact sky-blue by day; in candle-light, however, it was astonishingly changed, not having then any blue in it, but being what I called red, a colour which forms a striking contrast to blue. Not then doubting but that the change of colour would be equal to all, I

requested some of my friends to observe the phaenomenon; when I was surprised to find they all agreed, that the colour was not materially different from what it was by daylight, except my brother who saw it in the same light as myself. This observation clearly proved, that my vision was not like that of other persons. [11, pp29–30]

In his self-investigation Dalton employed an assistant of normal colour perception to correctly name colours for him as he recognised that he lacked any standard of comparison. Describing his confusions, Dalton said that under daylight reds appeared dark and even 'dark drabs' while crimson and dark blue woollen yarns appeared equivalent. But under candlelight the reds seemed much more vivid than by daylight; crimson lost its blue equivalence and became yellowish-red while scarlet especially became 'a superb colour'. Most dramatic, as might be expected from the encounter with the geranium, was the apparent change in pink between daylight and candlelight. Under daylight pink seemed light (or sky) blue while under candlelight it became yellow or yellowish-red. Green and red also appeared equivalent under daylight with the leaf of a laurel plant and red sealing wax appearing alike. Dalton unhesitatingly concluded 'that I see either red or green, or both, different from other people.' [11, p34] How differently he saw them is summarised below:

(a) In the solar spectrum what others called red, he called 'shade or defect of light'. [11, p31]

(b) While normal people saw in the spectrum red, orange, yellow, green, blue and purple, he saw only yellow, blue and 'purple'. And his purple was more a darker blue than a distinctly different colour. (Dalton used the terms 'purple' and 'violet' synonymously. Seeing almost no red, he would indeed receive similar impressions from violet and purple. As he himself said, he never would have suspected that purple was a compound of blue and red had he not read it in Newton. Hence he considered purple a spectral colour. However, he was not alone in this misconception as some uncritical *normal* observers also called the violet end of the spectrum purple.)

(c) His yellows seemed to include the red, orange, yellow and green of normal persons so that from red to green in the spectrum he saw essentially one colour, yellow, in varying intensities; brightest for orange and descending in brightness towards green. However, there was a clear change in hue between the (normal) green and blue.

(d) In all points where he differed from other persons in describing colours, the differences were much less under candlelight than by daylight, with red and scarlet especially becoming noticeably more vivid. Compared to normal observers his colour perception was markedly altered by candlelight whereas theirs was hardly changed. Hence, Dalton's colour perception improved under candlelight.

Dalton was never satisfied with empirical descriptions, although in

Plate 1 (Left) Subtractive colour mixing. Three colour filters, magenta, cyan and yellow, are placed in front of a beam of white light composed of red, orange, yellow, green, blue and violet light. Each filter removes certain colours from the beam by absorption. Together, all three absorb all the light giving a black impression. (Right) Additive colour mixing. Three spectral colours (i.e. colours of light), red, green and violet, are projected onto a screen. Overlapping the three colours (addition) gives white. Notice that addition of two colours here yields one of the subtractive primaries on the left. The corresponding subtractive filter, therefore, transmits those colours but absorbs the third.

Plate 2 Colour triangle of Thomas Young given in his *Course of Lectures on Natural Philosophy and the Mechanical Arts*. [Chapter 1, 14] According to Young, employing red, green and violet primaries, the triangle gives all the possible shades of colours.

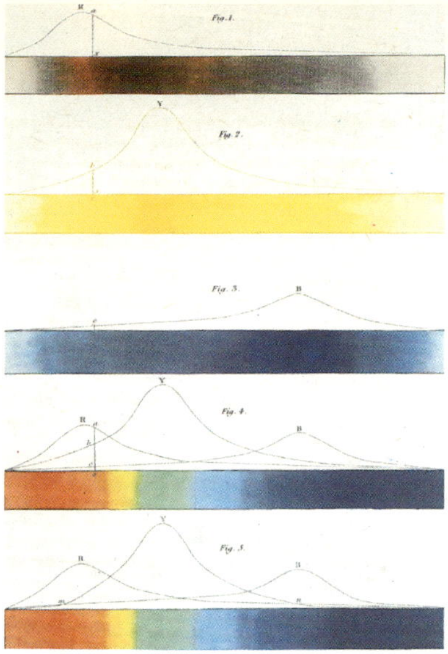

Plate 3 Brewster's triple spectrum as proposed in 1831. Three equally long, independent spectra exist for red, yellow and blue. They overlap as shown in the bottom figures and add at any point in the spectrum to yield all other spectral hues.

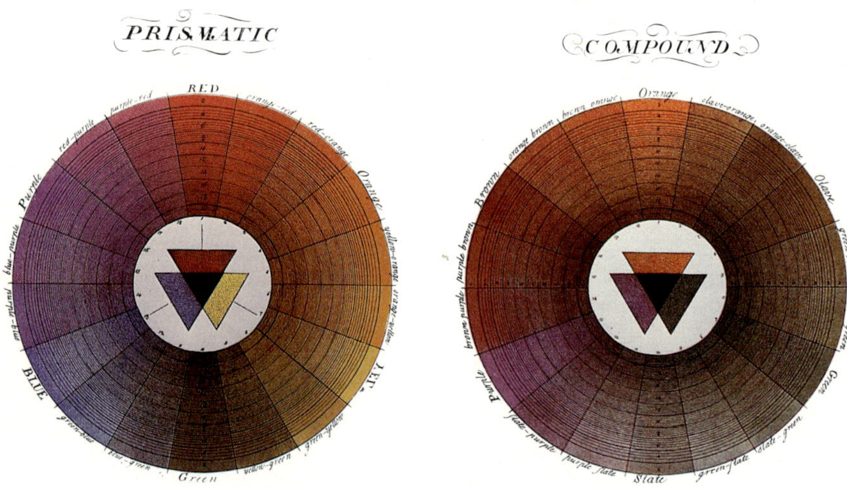

Plate 4 Colour circles of Moses Harris in 1766. The circle of prismatic colours is based on red, yellow and blue primaries whose additive property was (incorrectly) supposed to give black. The circle of compound colours uses three intermediate colours from the prismatic circle, orange, green and purple, as its primaries (colour restorations by Faber Birren 1963).

Plate 5 A colour triangle painted by Georg Cristoph Lichtenburg to illustrate Mayer's colour triangle. Lichtenburg used six transitions between the primaries instead of Mayer's proposed twelve. (Reproduced from *Tobias Mayer, Opera inedita: The First Translation of the Lichtenburg Edition* by Eric Forbes, by permission of Macmillan (London and Basingstoke).)

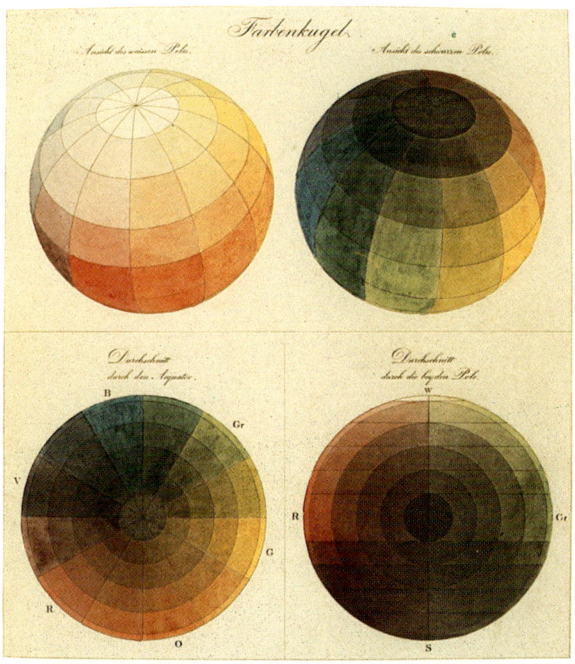

Plate 6 Colour reproduction of the colour sphere of Philip Otto Runge, 1810.

Plate 7 Contrasting colours produced by D R Hay in 1845 in *A Nomenclature of Colours, Hues, Tints and Shades*. [Chapter 4, 16] The upper hues are desaturated with white and contrasted with the hues below them darkened with black. The scheme in effect compares purity of colour with brightness of colour.

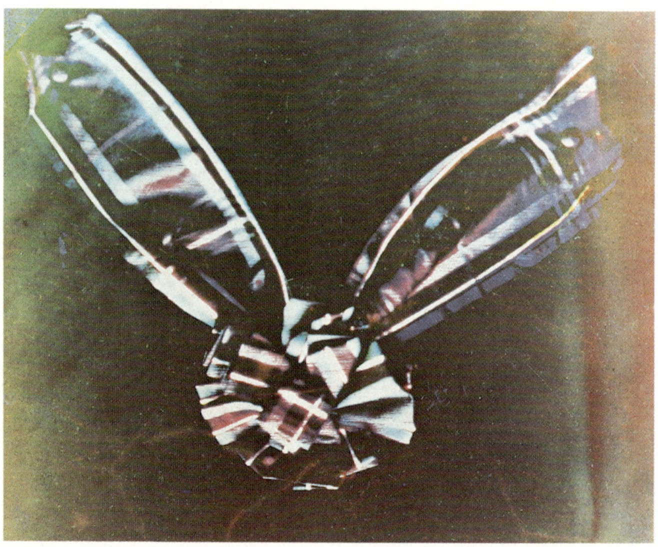

Plate 8 Maxwell's colour photograph proposed within the context of his first major paper on colour in 1855 and demonstrated at a Friday Evening Discourse at the Royal Institution in 1861. (Courtesy of the Science Museum, London).

identifying these symptoms so explicitly he had already done a great service. Searching for an explanation for his peculiar vision he observed that a sky-blue transparent liquid modified the light of a candle making it appear similar to daylight. It also gave pink, in daylight, the appearance of light blue. From this Dalton hypothesised that one of the humours of his eye was tinged with blue, in particular the vitreous humour. (Inspection of the outer eye revealed no bluish tinge.) This would explain the geranium observation. Considering pink (incorrectly) as a mixture of red and blue, the blue-tinged humour would absorb the red rays reflected from the flower but pass the blue rays thus giving it the blue appearance. For candlelight Dalton knew that the proportion of red and orange to blue was higher than in daylight. Therefore, the candlelit geranium would reflect higher ratios of red and orange compared to blue than in daylight. But since he could not see the red and presuming that the intensity of the orange was greater than that of the blue, the flower would appear a new colour to him under candlelight. It would be one of his shades of yellow. With one exception, Dalton was able to show that this hypothesis fitted all the facts concerning his colour vision. However, he could not understand 'why the greens should assume a bluish appearance . . .' [11, p44]

It is interesting to see within what tradition Dalton viewed his discovery. The scientific journals then often published, besides articles of bona-fide scientific interest, many odd curiosities were stated without comment. The *Philosophical Transactions*, for example, has reports on a two-headed calf, an extremely loud thunderclap, a deer's horn found in the heart of an oak, a fiery meteor, and many other wonderful oddities. It was all part of the new scientific spirit of careful observation as contrasted with earlier methods of generalisations from limited knowledge. The roots of the new philosophy lay partly in the explicit examples set by its founding fathers, such men as Gilbert, Galileo, Kepler, Harvey, etc, all of whom practised the new method but wrote little about it. Contrasted to them Francis Bacon, a non-practitioner of the new science, became its self-appointed populariser. For him the road to reality in science was inference from experience rather than the scholastic appeal to authority and rigorous deductions. Then after reaching sound principles by induction from numerous observations, deduction was also legitimate. Later scientists—and many dilletantes—regarded themselves as pursuing Bacon's philosophy although they hardly followed his prescription as rigorously as they supposed. To actively aid and foster this new spirit the Royal Society of London was founded in 1662 with a view to 'consider and discourse on philosophical experiments and observations' and to 'discourse upon, rarities of nature and art; and thereupon to consider, what may be deduced from them . . .' [13, p61] In this sense the Royal Society supported and solicited first-hand reports on all forms of natural knowledge without regard to underlying reasons. Following Bacon, the members argued that while the reasons currently remained hidden sufficient additional examples might lead to new knowledge. These seeming oddities then really represented part of a 'well developed form of empiricism, unique to the Royal Society, not by any means always perfectly consistent, but always successful, providing a model which others might follow.' [13, p57] The articles by Huddart, and Scott therefore, were part of the tradition of reporting odd curiosities which in this case proved to be important ones, for succeeding 'odd' reports on the same curious vision really did lead in a rather Baconian sense to an understanding of a new anomaly of vision.

The Royal Society of London however was hardly the only one of its kind. Similar societies sprang up all over Europe while for those away from the main centres of science, numerous provincial scientific societies consciously emulated the model of the Royal Society. [13, p75] The Literary and Philosophical Society of Manchester, of which Dalton was a member, was one such provincial society. In the best of traditions, its memoires also then carried curious reports together with papers of much sterner stuff. It was within this long tradition that Dalton originally viewed his peculiar vision. Having definitively investigated himself and his brother whom he discovered was also colour blind, Dalton stated 'I conceived the design of laying our case of vision before the public, apprehending it to be a singular one.' [11, pp37–8] However, having read the case of Harris he decided to investigate Harris' family further. Accordingly he sent some 20 specimens of coloured ribbons to a friend living near the Harrises with instructions to exhibit them to the survivors of the family. It turned out that there were six sons and one daughter of whom four sons were colour blind, apparently with symptoms similar to Dalton's. (Huddart had mentioned only two of the sons.) It was this act which set Dalton's evidence apart from the tradition of oddities. For as he himself said

> It then appeared to me probable, that a considerable number of individuals might be found whose vision differed from that of the generality, but at the same time agreed with my own. [11, pp38–9]

This realisation together with his actions following it lifted colour blindness out of the category of a singular occurrence. Dalton conducted a further search for similarly affected individuals finding three more out of 50 tested. [11, p39] He, moreover, reviewed the cases of his own family and that of Harris bringing the latter once again to public attention. This review served to demonstrate that his 'peculiar vision' was not at all singular but indeed fairly common. It was common, at least, among males, for Dalton remarked that he could not find a single female example. [11, p40] The review finally indicated that this defect in vision seemed to be hereditary since all his documented cases showed other members of the family similarly affected. Unfortunately he missed the case of Scott whose familial descriptions in this respect were quite convincing.

Dalton's search for and discovery of other colour blind people indicates that the historical difficulty of identifying colour defective vision rested on the problem of finding suitable subjects. The examples already mentioned illustrate the perplexity a colour blind individual experiences. This is certainly due to the difficulty of communication on colours between himself and colour normal persons. In general, we hardly suspect that colours which (correctly) appear different to one individual will appear equivalent to another. *Vice versa*, the colour blind person hardly suspects that any difference exists in colours which are obviously 'the same'. Colour blind people, of necessity, cannot agree with others on the meaning of colour words so no common language exists between them and the rest of the world; all is confusion. Therefore, one must expect quite some delay until the colour blind person recognises, if he ever does, that the confusion lies within himself and not in the language of others. Dalton summed this up succinctly by saying 'I was always of the opinion, though I might not often mention it, that several colours were injudiciously named.' [11, pp28–9] Recognition of other visual problems, especially those requiring lenses, certainly did not carry over to this more subtle problem.

There is, however, another difficulty the colour blind person experiences. Among those from whom he requests information about colours there is often marked disbelief that he doesn't really see colours exactly as they do. Children generally have learned their colours by the age of five, so colour confusion beyond then can be regarded as a sign of slow learning. The colour blind person is often tacitly adjudged as stupid, even yet, so Huddart's concern for Harris' intelligence was reasonably founded at the time. It follows that the historical lag in recognising colour blindness also stems from an insensitivity in others ascribing colour confusions to stupidity instead of perceiving it as a real handicap. How many colour blind artisans have been considered stupid for mixing up their colours?

On the other hand, having established one's intelligence the amazement takes other forms. Requests for colour information are greeted with laughter. As Dalton related, 'I have often seriously asked a person whether a flower was blue or pink, but was generally considered to be in jest.' [11, p29] After several such experiences there is retreat into reticence. One simply doesn't wish to appear abnormal nor be forced to join in false jest as Dalton said he often did as a young man; [14] and it is very tiresome to constantly name colours (usually wrong) for the diversion of disbelieving people. Thus afflicted individuals are hard to find because they generally keep the problem to themselves. Dalton, like so many colour blind people, spoke freely about his difficulty in his later years, reputation established, but not we may note when he was a young man. Then, as said before, he didn't often mention it. And the author may mention that among his own students, those who are colour blind are reluctant to admit it.

An attendant problem in the process of self-discovery in the eighteenth century and earlier was almost certainly the rather drab surroundings. People then were not exposed to such a variety of colour as now and so opportunities for self-recognition were limited. Clothing was often sombre and buildings were mostly white half-timbered or red brick with little variety. The main objects for colour observation must have been animals and flowers. In this respect Harris mentions cherries, Dalton's initial diagnosis was made with a geranium, while cases reported in the early nineteenth century also frequently mention inability to see cherries, berries or fruit on trees. Indeed, if this is so there would even be a seasonal factor, rendering short the period of the year with sufficient variety of colours to observe. Small wonder then that colour blindness was recognised so late. The gradual emergence of the problem as one of scientific value is paralleled by every colour blind person in his early life. The whole difficulty of discovery was charmingly summarised by Brewster in a parable: [15]

> If one of the two travellers who, in the fable of the chameleon, are made to quarrel about the colour of that singular animal, had happened to possess this defect of sight, they would have encountered at every step of their journey, new grounds of dissension, without the chance of finding an umpire who could pronounce a satisfactory decision. Under certain circumstances, indeed, the arbiter might set aside the opinions of both the disputants and render it necessary to appeal to some higher authority,
>
> to beg he'd tell them if he knew,
> whether the thing was *red* or *blue*.

As subsequent events will show, that higher authority ultimately became a series of important experiments by different investigators over a period of about 50 years.

In Dalton's pivotal paper there was ultimately one point of disagreement between himself and two other scientists of his generation interested in the problem. Neither Thomas Young nor John Herschel was satisfied with the hypothesis of a blue vitreous humour. Young pointed out in 1807 that Dalton's vision could have another explanation. Hypothesising three sets of colour receptors or nerves as part of his theory of colour vision, Young stated that Dalton's vision could be more simply understood by supposing the red receptor to be either absent or paralysed. [16] This would explain the one fact Dalton left unresolved. In the absence of a red receptor, a supposedly slightly higher response of the blue receptor over that of the green would indeed give greens a bluish tinge, even in candlelight which has a low blue content.

Young's explanation was rejected by Dalton—and not surprisingly, for Young's theory and hence his explanation were completely hypothetical while Dalton's blue humour inference had at least one good observational foundation. Dalton preferred a clear physical interpretation over unknown anatomical possibilities. Dalton wrote to John Herschel, probably the foremost optical expert of his time, about the question in 1833. Herschel having made an investigation of the subject himself replied to Dalton saying:

> It is clear to me that you and all others so affected perceive *as light* every ray, which others do. The retina *is excited* by every ray which reaches it, nor, so far as I can see, is there the slightest ground for believing that any ray is prevented from reaching it by the media of the eye. This, I know, is contrary to your opinion, but I speak advisedly and after careful consideration . . . The question, then is reduced to

> one of pure sensation. It seems to me, that we have three primary sensations when you have only two. [17]

Without committing himself to anatomy, Herschel agreed with Young at least on the outlines of the trichromatic response mechanism.

The question of the blue humour was finally settled by post-mortem examination of Dalton's eyes, at his own request. Both humours in both eyes were found clear and transparent. In one eye, the posterior part was removed leaving both humours in place and scarlet and green objects were then viewed through the eye. Placed next to each other both colours retained their distinct hues rather than appearing equivalent. [18, p202] So the blue humour hypothesis was demonstratively proven wrong while physiologists noted some years later that no such tinge had ever been seen in thousands of such examinations. Nevertheless Dalton's hypothesis continued to be discussed, if not accepted, as late as 1855 simply as one possibility among various explanations considered.

At the time of Dalton's death (1844), however, another more unusual opinion on colour vision than those we have seen so far was also struggling for existence from the field of phrenology. Phrenology was a system of psychology first formulated by F J Gall in 1798 in Austria and much developed by two followers in Britain, J C Spurzheim and George Combe. It proposed that mental powers and psychological attributes could be analysed into separate, independent faculties and that each faculty resided in a specific region on the *surface* of the brain. Since the size of each region naturally measured the extent of that faculty, it followed that a relationship must exist between the outer surface of the skull and the contours of the brain below it. Therefore, the extent of the various faculties were recognisable by examining the outer surface of the head. Phrenology so caught the public imagination that in 1832, for example, there were 29 phrenological societies in Great Britain alone and several journals of phrenology published in Britain and America.

Naturally phrenologists regarded colour vision as one faculty of interest. The colour discrimination faculty was assigned to an 'organ of colour' usually placed in the region of the frontal sinus where it would be exhibited by the superciliary ridges of the eyes. Large ridges meant good colour vision, small ridges, poor colour vision. Since the 'science' held such fascination, serious writers on topics of interest to phrenologists were forced in effect to give it due attention if only to refute their position. This was the case at Dalton's post-mortem for his famous vision was especially espoused by the phrenologists as justifying their ideas.

Besides the dissection of Dalton's eyes, a phrenologist was allowed to take casts of his skull and brain to uphold the phrenological position on Dalton's colour blindness. Although the physician in charge could not advocate his opinion he felt honour bound to report the results saying,

> Mr Bally (formerly assistant to the late Dr Spurzheim) was present, and took casts of several parts of the brain, and the cavity of the skull. He pointed out a remarkable prominence on the frontal portion of the

orbiter plate (which represents the phrenological site of the 'organ of colour'), and the imperfect or deficient development of the convolutions of the anterior lobes, which rested upon them. Of course, Mr Bally adopted this as the true explanation of the peculiarity of Dalton's vision; and we as witnesses (not only without faith in 'phrenology', but even opponents) are bound to record the facts. [18, p201–2]

The phrenologist's opinion on Dalton was most definitively refuted, however, eleven years later by George Wilson, Regius Professor of Technology at Edinburgh University, in a detailed work on colour blindness to be discussed later. Essentially, he noted that portraits of Dalton showed rather large eyebrows which was inconsistent with phrenological opinion. Yet phrenology at the time was taken seriously and writers on colour perception often included the phrenological opinion when discussing possible causes of colour blindness.

Young's explanation of Dalton's vision is in fact nearer the correct one. Spectral sensitivity curves for normal vision and vision like that of Dalton are shown in Figure 7.1. [19] The curve for the protonope, into which class Dalton (and very likely Scott) would now be placed, shows that there is a much reduced sensitivity to the red end of the spectrum. The threshold of vision is shifted markedly toward the orange and we can readily understand why for Dalton red objects appeared shadowy or black. Suffice it to say here that Dalton's self-diagnosis together with clear evidence for the frequency of other cases marks the real beginning of investigation of colour blindness. The problem was finally recognised.

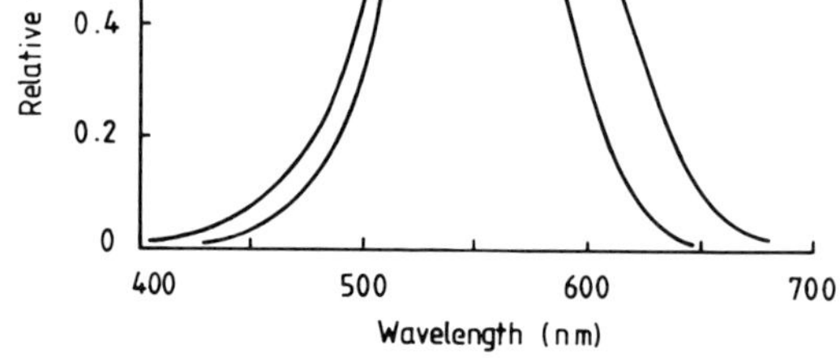

Figure 7.1 Spectral sensitivity curves for a normal subject and a protanope.

If the problem was recognised it was also ignored. With a few exceptions there was a hiatus of over 20 years between Dalton's report and any serious new discussion of abnormal colour perception. John Wardrop, for example, an ophthalmologist, mentioned it in 1808 in a medical textbook but gave no details beyond those already known. [20] Goethe too in his *Colour Theory* of 1810 described two cases of colour blindness saying already that it is found in many persons. Interestingly he called attention to the communications difficulty stating:

> If anyone, accidentally conversing with these individuals, happened to question them about surrounding objects, their answers occasioned the greatest perplexity, and the interrogator began to fancy his own wits were out of order. [21]

By simply asking his subjects to name colours Goethe noted that they 'apparently' saw yellow, reddish-yellow, and yellowish-red the same as normal persons. A pale crimson was considered the same colour as the sky and called blue. So too was a rose called blue. In general they confused rose colour, light blue and violet; and confused green with dark orange and red-brown.

Goethe was most perplexed at the confusion of sky and rose colour. Since the sky and rose were interchangeably called red or blue, notwithstanding the first observation, the question was: Did they see the sky and rose both blue or both rose colour? With no objective answer possible, Goethe assumed one. In his colour theory, orange = red + yellow; green = yellow + blue; violet = red + blue. If no blue were visible to these observers, violet and green would also be absent so they would see only three colours, red, orange and yellow. Therefore the sky and rose would both be seen as red regardless of the sky ever having been named blue. Goethe simply assumed that his subjects saw both the sky and rose as red without further attempts at differentiating their responses. For Goethe therefore colour blindness was an insensibility to blue resulting in a world viewed with a reddish-orange tinge. On this basis Goethe named their deviant vision 'acyanoblepsia' denoting a non-perception of blue. However, had he tested his subjects with a spectrum as Dalton had done with himself he would have found that they did see blue for their symptoms, so far as one can determine, are similar to Dalton's although less severe. Goethe's colour theory, having a vicious anti-Newton bias, was generally ignored by the scientific community so these naive ideas ultimately played little role in succeeding events beyond adding two new cases to the slowly growing list. Goethe in any case was the first to honour the new phenomenon with a name in print albeit one misconceived.

The hiatus following Dalton's paper ended in the 1820s with a flurry of new cases. Since they are very similar we will note just a few in order to appreciate their style and the limited understanding of abnormal colour vision at the time.

One of the earliest cases of colour blindness carefully reported after Dalton was an account by John Butter, a physician, in 1822. [22, pp135–41] A young man was found to see, similarly to Dalton, red and green as identical colours, blue and pink as identical colours but different from the red and green, while indigo and violet were both described as purple. Beyond the details of the confusions Butter then considered the possible causes of this defective vision. Noting that the visual acuity of his subject was otherwise particularly good and his eyes were free from other difficulties, Butter concluded that the cause must be a physiological one and not ascribable to the optics of the eye. He presumed the problem to lie in the brain which 'feels all the sensations of light, sound, odour, and taste' [22, p139] for the peculiar vision certainly did not result from the mechanical construction of the eyes. Brewster, as editor of the journal in which the paper appeared, saw fit to comment favourably on this conclusion. He pointed out Young's earlier hypothesis as contrasted with that of Dalton stating that although there was no evidence for Young's 'fibres' it was sufficient to suppose the (abnormal) retina to be insensitive to certain colours. Without elaborating on that cause he pointed to an earlier work of Wollaston's on hearing by way of analogy. Wollaston had shown in 1820 that not everyone heard equal ranges of frequencies. Some persons with acute hearing became insensible to sound above and below certain musical notes while others continued to hear beyond those limits. Wollaston ascribed this to differences in the tympanum and Brewster now ascribed a similar insensitivity in the retina to certain colours of light:

> if a sound ear, therefore, may be deaf to sounds of a certain pitch, without our looking for the cause of this in the form of any part of the brain, why should we appeal to such an uncertain guide for an explanation of the analogous phenomenon of the insensibility of the eye to certain colours? [22, p141]

Brewster rejected Butter's interpretation and accepted a modified version of Young's theory without the fibres.

That year also saw two more cases reported, again in Britain, by a physician, Whitlock Nicholl. [23] One was a man who, like Dalton, could not distinguish green from red; he called grass red and could not distinguish red fruit on a tree from its leaves. He could, however, apparently distinguish yellows and blues while crimson curtains appeared to him blue by day but red by candlelight. The other, a child of eleven, never used the word 'green'. Dark green was called brown, pale green was called light red, dark yellow and light brown were called red, light red and rose-colour were called light blue. Red, however, was correctly called red while various blue shades were all named one colour—blue. In addition the child saw only red, yellow and purple (i.e. violet) in the spectrum although Nicholl did not so test his adult subject. Nicholl's means of testing was only to ask his observers for names of colours comparing their names with his own. He was unaware of linguistic differences necessarily arising between himself and his observers and simply accepted their statements on face value. If nothing else, for example, we have seen more sensitive observers name grass green simply by comparison of their vocabularies with other people. Scott, Harris' brother the sailing master, and Dalton all said in this respect that they took their idea of green from grass even though they could not discriminate between red and green. Therefore they correctly named grass as green while Nicholl's cases incorrectly named it red. It is difficult, then, to determine from Nicholl's antiquarian descriptions where the difficulties lay especially for the child, although they clearly lay in the red–green end of the spectrum. Nicholl, in a separate article on their difficulties, illustrated his own confusion. [24] In a tortuous analysis he concluded that they could not see blue; while not mentioning Goethe he invoked essentially the same theory and assumptions to arrive at the same incorrect conclusion.

Compared to Nicholl's reports, a very well described case of colour blindness was reported by George Harvey in 1826. [25] Here he described a tailor whose colour vision had led him, for example, to repair a dark-blue coat with a patch of crimson, and a black dress with crimson. Harvey asked him to name colours, to arrange five pieces of cloth in white and grey by their shades, and to bring him 8 or 10 specimens of green cloth which lay in the shop. In naming colours, the tailor confused dark-blue with crimson and black, purple with blue, light green with dark-yellow and named red as blue. He could arrange the white and grey cloths but could not identify the green cloths he was requested to bring. In fact, he brought some of red-black and black-brown as well, unaware of any difference. Most interesting, however, was Harvey's explanation for his subject. In an oblique allusion to Young he stated

> Of the hypotheses that have been proposed to account for the phenomena in question, that which refers it to an insensibility to certain impressions of light, appears at once the most simple and philosophical. [25, p261]

Harvey's paper was seen as important enough to be translated into French and printed in its entirety in the *Bibliothèque Universelle* for 1827 where it received wide continental attention. Pierre Prévost, the physicist–philosopher, for example, saw fit to comment on it because he thought Harvey's article was deficient for not comparing his subject with Dalton or other known cases. Referring to Harvey's tailor, Prévost wrote that he 'appears to differ from the number of those whom I am accustomed to call *Daltonians* only by a slight degree of darkness in the shade.' [26, p321] And later Prévost stated, 'I do not hesitate to pronounce him a Daltonian.' [26, p322] Although Prévost reported on no work of his own (and apparently did none on this subject), his remark does indicate that the subject was at least well known to him. Moreover, the noun 'Daltonian' for a colour blind person had apparently been in use for some time by then having definitely been used in oral instruction in Geneva since about 1800. [27] However, Prévost has the dubious distinction of being the first to use the term in print. Dalton had unwittingly contributed his name in Europe to a curious effect.

Brewster brought defective colour vision to public notice again in 1826 in a popular review of the cases of Harris, Scott, Dalton, Butter and Harvey while noting in passing yet another three. Still searching for reasons he dismissed Dalton's view as mere conjecture, while Young's hypothesis was simply stated without comment. Noting his previous analogy with Wollaston he expanded on that saying:

> If we suppose, what we think will ultimately be demonstrated, that the choroid coat is essential to vision, we may ascribe the loss of red light in certain eyes to the retina itself having a blue tint. If this should be the case, the light which falls upon the choroid coat will be deprived of its red rays, by the absorptive power of the blue retina, and consequently the impression conveyed back to the retina, by the choroid coat, will not contain that of red light. [28]

Thus to account for loss of vision beyond certain limits in the sense of loss of hearing Brewster invoked a mechanism of selective absorption, an idea certainly influenced by his earlier work on absorption. Whatever the merits of this idea, more like Dalton's now than Young's, Brewster must be viewed as an important contributor to knowledge of colour blindness in that he published as editor of two respected journals several articles on colour blindness together with commentaries. Considering the paucity of work in the field, this served the purpose of keeping the subject alive and before the scientific community. In fact a casual perusal of the article indicated at least eight current cases of defective colour vision which together with his references totalled some 15 cases in Britain alone reported since 1777. Finally Brewster showed in this article and another in 1829 [29] that there were at least three hypotheses by which to account for the known observations. Together with a similar number of cases reported on the Continent, this peculiar vision was beginning to look less rare.

A very thorough series of tests on a colour blind subject was reported by Sir John Herschel in 1827 in his *Treatise on Light*. [30] They are important in documenting yet another case and in their mode of testing. While other investigators presented dyed cloths or coloured glasses and paints to their subjects, Herschel employed only spectral colours. Using polarised light Herschel could present either individual spectral colours or compound optical colours. The subject, viewing two coloured circles, was then required to name or describe the hues. The results of these observations Herschel reported as in Table 7.1 in which the subject's descriptions were compared with a normal observer. [30, p435] Of this Herschel stated:

> We have examined with some attention a very eminent optician, whose eyes (or rather eye, having lost the sight of one by an accident) have this curious peculiarity, and have satisfied ourselves, contrary to the received opinion, that all the prismatic rays have the power of exciting and affecting them with the sensation of light, and producing distinct vision, so that the defect arises from no insensibility of the retina to rays of any particular refrangibility, nor to any colouring matter in the humours of the eye, preventing certain rays from reaching the retina, (as has been ingeniously supposed) but from a defect in the sensorium, by which it is rendered incapable of appreciating exactly those differences between rays on which their colour depends. [30, p434]

Table 7.1. Observations of the spectrum by a colourblind observer compared with a normal observer as given by Herschel.

Colours according to the judgment of an ordinary eye		Colours as named by the individual in question	
Circle to the left	Circle to the right	Circle to the left	Circle to the right
Pale green.	Pale pink.	Both alike, no more colour in them than in the cloudy sky out of window.	
Dirty white.	Ditto, both alike.	Both darker than before, but no colour.	
Fine bright pink.	Fine green, a little verging on bluish.	Very pale tinge of blue.	Very pale tinge of blue.
White.	White.	Yellow.	Blue.
The limit of pink and red.		Both more coloured than before.	
Rich grass green.	Rich crimson.	Yellow.	Blue.
		Better, but neither full colours.	
Dull greenish blue.	Pale brick red.	Blue.	Yellow.
		Neither so rich colours as the last.	
Purple (rather pale.)	Pale yellow.	Blue.	Yellow.
		Coming up to good colours, the yellow a better colour than a gilt picture-frame.	
Fine pink.	Fine green.	Yellow, but has got a good deal of blue in it.	Blue, but has a good deal of yellow in it.
Fine yellow.	Purple.	Good yellow.	Good blue.
		Better colours than any yet seen.	
Yellowish green.	Fine crimson.	Yellow, but has a good deal of blue.	Blue, but has a good deal of yellow.
Good blue, verging to indigo.	Yellow, verging to orange.	Blue.	Yellow.
		Both gay colours, particularly the yellow to the right.	
Red, or very ruddy pink.	Very pale greenish blue, almost white.	Yellow.	Blue.
Rich yellow.	Full blue.	Fine bright yellow.	Pretty good blue.
White.	Fiery orange.	Has very little colour.	Yellow, but a different yellow, it is a blood-looking yellow.
Dark purple.	White.	A dim blue, wants light.	White, with a dash of yellow and blue.
Dull orange red.	White.	Yellow.	White, with blue and yellow in it.
White.	Dull dirty olive.	White.	Dark.
Very dark purple.	White.	Dark.	White.

From these results together with another series of colour-naming experiments Herschel came to the important conclusion that

> It appears by this, that the eyes of the individual in question are only capable of fully appreciating blue and yellow tints, and that these names uniformly correspond, in his nomenclature, to the more and less refrangible rays, generally; all which belong to the former, indifferently, exciting a sense of 'blueness' and to the latter of 'yellowness'. [30, p435]

What is not clear from his table of colours (normal eye) is the red extent of his spectrum. Certainly his subject experienced some red response but whether it extended to the full red dimensions of the spectrum has not really been established here; so it is doubtful whether Herschel was justified in concluding that all the prismatic rays could evoke a sensation of light. It was simply all the rays produced by his apparatus. In fact the constant reference by his subject to yellow and blue reminds one of Dalton who saw only yellow, blue and 'purple' in the spectrum and Dalton certainly had a sharply reduced red sensitivity. Therefore Herschel's conclusion on this point is questionable and his rejection of (Young's) insensitivity of the retina to a particular colour also questionable. He also rejected Dalton's hypothesis and it was upon these experiments that he had based his reply to Dalton in 1833. The reference to the 'sensorium' was influenced by physiological speculations on the brain as the interpretor of the senses. Lacking detailed knowledge it was reasonable to say that some portion of the brain, as yet unknown, was responsible for the abnormal vision. Not then recognising the retina as an extension of the brain itself, his rationale for speculation was as good as anyone else's. In any case, an insensibility of the retina and a defect of the sensorium preventing appreciation of colour differences are very similar.

Herschel's results were important for another reason in particular. His *Treatise on Light* was translated into French and German and, coming from the authoritative pen of Herschel, they demanded and received wide attention adding yet another case history to the list as well as a new experimental method. Moreover, Herschel's discussion of colour blindness led directly in his *Treatise* to a discussion of Mayer's three-fold classification of colours and Young's choice of primaries; it was a hint that information on these subjects might be forthcoming from data on colour defective people.

Yet two more cases were reported in 1829 again by a physician, Hugh Colquhoun. [31] One was a self-description solicited from an acquaintance of Colquhoun while the other was of an orchid grower he personally interviewed. The first case reported in rather contradictory fashion that he could distinguish 'the brilliant reds and greens in the plumages of certain birds . . . with perfect accuracy' [31, p13] while then saying 'the colours between which I have the most difficulty in distinguishing the difference are . . . those reds which approach nearest to scarlet and the yellower greens such as the leaves of the lime tree, the birch, the elm, etc.' [31, p13] More significantly he could not discriminate red sealing wax from grass nor a scarlet cloth hung on a hedge; in fact he once collected some bright green lichen from the roof of a hut thinking it the same 'brilliant' red as the roof itself. He was much impressed by the 'scarlet' lichen! How he could distinguish red and green plumages, therefore, with perfect accuracy even he could not understand. Possibly, brighter viewing conditions might evoke stronger responses but his meaning of 'brilliant reds and greens' in birds is hard to judge. In other respects darker shades of green, blue and purple were distinguishable but pale shades of green, pink and pale blue were all alike; also crimson and pale blue were equivalent colours which resulted in his being the butt of a joke by some acquaintances. Upon learning of his difficulties, a woman left the room, removed her rouge and replaced it with a blue pigment to everyone's amusement except his own for he saw no difference until the joke was explained to him. Almost literally his friends played a joke on a blind man.

Colquhoun's second case was a man who, like so many, first noticed his difficulty with leaves and fruits. Unfortunately his later livelihood depended on colour discrimination. At first a weaver he was forced to leave that trade because of constant mistakes in selecting threads for the patterns; on cloudy days in particular red, black, green and purple threads were all identical. Alternatively he settled on orchid raising where however his problems still bothered him. Colquhoun tested this man carefully asking for colour names of pieces of paper, colours from Symes' *Werner's Nomenclature of Colours*, and of surrounding objects. The resulting description of his problem is very detailed although it is fairly well summarised by the man's own history as a weaver. Basically he could not distinguish between reds and greens, darker reds appeared more like mud colour or brown, even bright greens seemed 'dull drabs', blue was usually correctly named but dark blues were sometimes called black, and purple was always confused with blue but distinguished from other colours. At dusk his colour perception worsened with his responses becoming even more uncertain. Finally under candlelight he could discriminate orange but not reds although carmine was called 'red'. What happened to green is left unsaid. As might be expected from Dalton's experience, blues changed their appearance so there was now confusion between pale blue and pink while all dark blues appeared black. In general Colquhoun reported quite thoroughly on the person he actually tested but he depended entirely on autobiography for the details of his other case. He did not appreciate any difference in procedure between personal investigation and solicited statement. Nor did he see fit to compare his cases with any of the others then known but he did give a rather complete listing of them all. Unlike his predecessors, however, he refrained from speculating on causes by pompously warning 'that it is *facts*, and not *reasoning* of which there is the greatest scarcity in the philosophical world, when any difficulty is to be solved.' [31, p21] Probably he was right for there were already any number of reasons available without any means of choosing between them. The field was just then new enough to really require an accumulation of case histories and real diagnosis before decisions regarding causes could be made.

These case histories indicate the form in which descriptions were usually reported and the variable quality of observations. Dealing with only one or two subjects at a time they were all limited in scope, were often superficial and sometimes contained second-hand material. They had an antiquarian quality which hardly differed from Huddart's original article in 1777. Except for Dalton nobody was interested in frequency of occurrence although there was casual recognition that the phenomenon is not so rare. References to family members similarly affected were sometimes included while just as often they were not; again Dalton's remarks on this point together with Scott's were still the best. Some 30 years after Dalton's work his was still the best of all published reports. So although the field was accumulating fresh examples they were all isolated cases of different times and places ranging from Germany to Great Britain. Interestingly, however, almost all new cases were reported in Britain.

Beyond this, the method of testing was always the same. Using dyed ribbons, wools, cloths, papers (and any surrounding objects) testing consisted only of asking for colour names. This in turn evoked a degree of linguistic confusion which revealed colour discrimination problems and also made their nature difficult to judge. Harvey had partly avoided this communications gap by asking his subject to select examples of coloured cloth but he also asked for colour names as well. Worse, considering the odd bits and pieces of coloured materials used there was no consistency to the actual colours employed which could only add to the attendant difficulties. Only Herschel's apparatus brought a degree of reproducibility to the colours exhibited but he too resorted to asking for colour names. From Huddart to Colquhoun in 1829 colour naming was the only testing approach followed, being an obvious way to proceed. For almost the next decade this remained essentially the style of all new investigations and the 1830s saw as many or more new cases of colour blindness reported as the previous decade. By then, however, it was obvious that abnormal colour vision was no rarity; yet beyond merely accumulating a catalogue of similar confusions little new was being learned. An original approach was definitely needed.

In view of the linguistic difficulties already mentioned no investigator could really rely upon the names of colours supplied by his subject. This procedure tended to mask the nature of the defect, and worse, it could even give wrong impressions of the abnormal vision. One can easily see, for example, from Herschel's comparative namings whether a person has defective colour perception, but it is difficult to appraise the colour sense lacking and its severity. Obtaining at best only incomplete data, such a procedure can only identify gross colour recognition difficulties. It can never make a differential diagnosis as to the specific type of defect. None of the writers to this time even suspected the occurrence of different types or degrees of colour blindness. As the case histories indicate all subjects were simply lumped together and described in rather naive, antiquarian terms. These logical flaws were noted by August Seebeck who in 1837 produced an important new approach to colour identification by colour blind persons. [32]

Dalton had found several of his new subjects among his own pupils. Seebeck apparently took up this hint for he searched for subjects, captive as it were, in the two top classes of a Gymnasium. Here he found that of 41 students five had defective colour vision thus illustrating the frequent occurrence of the problem. [32, p178] These together with eight other cases plus one unpublished case of his father's constituted Seebeck's group of subjects in whom he diagnosed not only different categories of colour blindness but also differing magnitudes of error within each category.

Seebeck's diagnostic method was simple in conception but definitely arduous in execution. Under sunlight he presented an unordered assortment of 300 coloured papers to each individual, who was then requested to sort them into a series. By asking for an ordering process communication difficulties between the subject and Seebeck were overcome while he could see from the ordering (or misordering) the colours confused by his subjects both as to magnitude and type. But the subject was faced with a long and detailed trial of his colour perception.

Besides ordering of coloured papers, Seebeck also used prismatic colours and interference colours. But of these modes of investigation Seebeck noted that while they provided colours of known optical content, the experimenter must rely upon the erroneous naming of colours [32, p177] a procedure he was essentially trying to avoid. In general he first requested the paper ordering process, and then having a knowledge of the person's difficulties he confirmed these with spectral and interference colours. Tests with the spectrum produced by a prism and cast on a screen were particularly useful in determining the regions of maximum and minimum sensitivity of his subjects to specific colours. He also determined their limits of sensibility by having them draw lines at various parts of the spectrum and then comparing these positions with those drawn by himself as a control. From this he discovered that six of his fourteen subjects could hardly see red at all, while the other eight, although definitely abnormal, could see the full extent of red in the spectrum. This procedure, however, was not used as the primary diagnostic test.

Unfortunately, Seebeck does not state whether he asked for a specific ordering of the papers, but since he compared results from the papers with observations on the spectrum, we may guess that he asked for a sequence of colours corresponding to the spectrum. However, 300 papers would be more than ample in this respect so he probably also asked for suites of particular colours in terms of shades such as from dark to light. This would then allow for differences in sensitivity to brightness in addition to distinguishing the colours confused. Then a person who like Dalton saw red and green as equivalent would place red and green papers together, probably all the dark reds and greens together, and all the light reds and greens together. But another person, differently affected, might distinguish all the major classes of colours yet still confuse shadings of colours within any one group. For example, a very pale red might appear the same as a pale green although the subject could distinguish saturated reds and greens. Such a procedure would enable a differential diagnosis to be made in terms of brightness if such differences between individuals existed.

By carefully noting the mistakes of each subject, by intercomparing the orderings of different subjects, and by eliciting criticisms of one subject's ordering from another, Seebeck concluded that there were at least two classes of colour defective vision. Class one subjects were characterised as having

> a very deficient sense for the perception of all colours; it is the most imperfect for red and, which is necessarily linked with it, for the complementary green. These two colours are little or not at all distinguished from grey. Furthermore, for blue which they cannot distinguish too well from grey either. Their sense is best developed for yellow but also this colour looks to them much less different from the colourless than to the normal eye. [32, p221]

Persons in class two were similar to those in class one in many of their colour confusions but were markedly worse in red and blue confusions. Moreover, from their spectral observations Seebeck definitely determined that 'they have, which is not true for the first class, only a weak perception of the least refracted rays.' [32, p222] Thus he distinctly identified the protonopes (not his term) as his second class into which category he placed Dalton. Discovery of more than one type of anomalous colour vision was a distinctly new contribution to the young science.

An interesting confirmation of his classification was devised by Seebeck. Noting that class one was similar to class two in many respects but differed from it in being able to see red, he thought that class one people might be, in effect, converted to class two if red light were removed. Hence, he tested some members of his first class at dusk when red light would be of low level or even absent. He gave one member of class one the coloured papers to order and found that his mistakes differed from those made under full sunlight. They were now similar to his class two subjects under full daylight. Therefore, class one people were 'converted' to class two and this confirmed his finding that their main difference lay in the degree of red perception.

Beyond this larger classification Seebeck also noted errors within each class referring to brightness. This he confirmed immediately by having his subjects order (and name) coloured glasses held against the sky for comparison with their paper orderings. Here the colours were noticeably brighter than the pigment colours of the papers, his primary diagnostic method. Of two class one members, the ability to make finer colour distinctions was markedly improved for one but was minimal for the other. Similarly for two subjects in class two, who generally agreed on the paper ordering, subtle differences between various glasses were noticed much better by one than the other. Thus while they were both definitely class two members by the primary diagnostic means, two formerly 'equivalent' colours viewed under the new conditions became noticeably different for one person but not the other. Referring to these individuals Seebeck said,

> this shows that N possesses the deficiency to a lesser degree than H. The eye of the latter one is weaker on the whole, since the former one noticed a difference in the lively colours, but was lacking such a faculty to the same, or almost the same degree as the latter one when looking at the much weaker opaque pigments [32, p213]

Similar judgements were arrived at by criticisms of one class's paper ordering by the other class. As expected, there were different opinions on gross ordering but even when the two classes agreed on similar colours they disagreed on shadings within colour sequences. These observations led Seebeck to suspect that a wide range of colour defective vision might exist in these terms and indefatigable investigator that he was, he set out to find it. Submitting about 60 papers to various people he did not suspect of having colour defective vision he noted a few who misordered several papers and misnamed several colours. Thus there seemed to be cases where the difficulty was present in such a small but still noticeable degree that it formed a transition to the normal eye. [32, pp217–18] This he did not pursue however.

Considering all his cases together with that of a colour blind woman Seebeck concluded that he had observed a complete sequence of colour vision from the normal to the most deficient. In these conclusions he anticipated later work of Rayleigh (1881) in discovering exactly those people, now known as anomolous trichromats, whose colour vision depends on three colour variables but which is nevertheless abnormal. Exactly as Seebeck said, their colour vision ranges from the near normal to the markedly abnormal. Suffice it to say here that Seebeck certainly observed these persons as early as 1837 clearly recognising them for what they represented. Unfortunately the emerging science of colour vision was not yet ready to appreciate the significance of these remarks. Indeed it lacked the very methods to pursue Seebeck's deep insight.

It was Seebeck who performed the first really systematic objective investigation of colour blindness. While Dalton had brought the problem to public attention, investigations after him but prior to Seebeck were largely haphazard, performed on single individuals usually discovered by accident, and always by asking for names of colours. Seebeck pioneered a new approach to the subject, emphasised the frequency of occurrence of such people and demonstrated the inconclusive nature of previous investigations. He even reviewed all of the existing literature and from the mistakes described he was able to classify many of those people into his two categories. Some, however, were too poorly described to permit retrospective diagnosis. Not least of all his case histories, by considering family members so affected, demonstrated once more the hereditary nature of the problem. He actually tested the colour blind grandfather of one of his subjects and showed that both grandfather and grandson belonged to the same class, differing only in the magnitude of their mistakes. And, uniquely, this work led Seebeck ultimately to a colour blind woman which even Dalton had noted was apparently a rarity. With Scott's sister this would be only the second case noted since 1778 not counting a dubious possibility in between. But between 1777 (Huddart's descriptions) and now, 1837, many male examples were known. Seebeck tested the woman by the ordering of papers and found that she made the mistakes of his class one. In addition, active person that he was, Seebeck found her father to be colour blind, also in class one, a situation now known to be required for female colour blindness. His case of female colour blindness is certainly the first one well documented by a valid test procedure.

One thing only Seebeck had in common with his predecessors. Like them he too searched for some causal hypothesis for colour blindness. Noting Dalton's explanation he contrasted it with Brewster's analogy with hearing saying that others deduce it directly from a lack of sensitivity of the light nerves. [32, p229] This he had thought at first was a reasonable hypothesis. However, upon recognising his two classes he could not reconcile this hypothesis with two categories of mistakes as well as with varying magnitudes of mistakes within each class. Without rejecting the idea outright he did find the hypothesis insufficient for his data. Ultimately he found no satisfactory explanation and he simply said that the defect must be understood as frequently occurring and his test subjects were extreme cases of an imperfection which happens often to a lesser degree without being noticed. [32, p231]

All these developments notwithstanding, appreciation of colour blindness remained limited and research on the subject still relatively little. In 1830, for example, it merited only one paragraph in a 79 page article on physiology in the *Edinburgh Encyclopaedia* of which none other than Brewster was editor. William Mackenzie, one of Britain's leading ophthalmologists, devoted all of three pages to colour blindness in 1830 in his work *A Practical Treatise on the Diseases of the Eye*. While citing all the references, he mainly discussed the various hypotheses, not symptoms, of which he noted Dalton, Young, Brewster and the phrenologists' opinion regarding the organ of vision. Of these only Dalton's was refuted based on his experience with post-mortems while the others were simply stated without comment. Still it is significant that a textbook should include some discussion of what was certainly a very new discovery. A second edition of Wardrop's text *The Morbid Anatomy of the Eye* appeared in 1834 in which he simply lumped together all the symptoms of the few cases known to him. In a two page survey he ascribed the problem to a greater insensibility of the retina to the impressions of the blue- and yellow-making rays which he vaguely attributed to differing refraction of the colours through the humours of the eye. There was little to be learned from him. In 1840 the third edition of Mackenzie's book still contained essentially the same discussion as the first edition ten years earlier except that Herschel's opinion was added to the others and incorrectly counted as agreeing with phrenology. Seebeck's work was unknown to Mackenzie. In 1840 also appeared the second volume of Johannes Müller's authoritative work *Manual of Human Physiology* where he devoted exactly one page to colour blindness, simply describing Seebeck's results in terms of his two classes but missing the point about confusions due to brightness. The subject had rated more space in the 1792 edition of the *Encyclopaedia Britannica*. An 1849 work by Mackenzie, *The Physiology of Vision*, said nothing about colour vision while a survey of Continental physiologists and ophthalmologists shows a similar pattern. The inclusion of colour blindness in Mackenzie's book on diseases of the eye and its omission from a work on the physiology of the eye is indicative of its place in the thinking of medical researchers. As of 1840 the best work had been provided by two physicists, Herschel and Seebeck, and a chemist, Dalton.

The next major research in colour blindness came in 1843 from Élie Wartmann, Professor of Natural Philosophy at the Academy of Lausanne. [27] Coming once again from a physicist, it was clear that the issue was really one for the physical sciences. It was Wartmann's opinion after surveying the field that 'Daltonism' still hardly existed as a worthy scientific problem and this is certainly so. Unfortunately his own research skill hardly advanced the field but in its wealth of references and detailed discussion of hypotheses his writing proved to be very influential. His own investigations consisted of three subjects only and of these he saw fit to give a detailed report on only one. To be sure, the investigations judged by this singular report were detailed but they were more in the style of the predecessors of Seebeck. Thus Wartmann's results were subject to the logical flaw pointed out by Seebeck and contained no new discovery. At first Wartmann employed Seebeck's method of arranging coloured papers; but his papers were too glossy resulting in a glare which tended to mask the results. Therefore, he gave up this method, the best at the time, and resorted to the traditional one of asking for colour names. This he did by two means: in one he asked for names of the colours in a spectrum projected on a screen; in the other he asked for colours of glass plates seen by transmission when held against the sky. These two forms of observation were instituted to see whether the subject exhibited the same defect when viewing colours by reflected light (from the screen) and by refracted light (through the glass). [27, p174] This procedure seems unwarranted on at least two points. First it implies that the refracted light would be somehow modified as compared with that seen by reflection to give a different sensation. There was nothing in optical theory at the time to lead to such a suspicion. Rather quite the opposite. Neither the undulatory theory, by now generally well accepted, nor the emanation theory predicted any changes in the physical properties of light undergoing refraction or reflection. Secondly holding the coloured glasses to the sky produced colours essentially by absorption. Since Wartmann speaks of glass plates, presumably they had faces reasonably plane-parallel so that refraction through them would be a secondary factor compared to the absorption mechanism. In that there was already a good theory of absorption by Herschel in his *Treatise on Light*; there was also Brewster's triple spectrum which in 1840 had to be viewed seriously. But absorption by the glass plates was not mentioned by Wartmann nor was Brewster's theory. We are not surprised then to find that the 'absorption colours' showed 'the evident confirmation of the results obtained with the solar spectrum.' [27, p176]

Wartmann's other observations were essentially a repetition of Herschel's with polarised light, where again the subject was asked to name (or misname) colours. From this procedure and the one with coloured glasses, Wartmann constructed long tables of misnamed colours which show an obvious defect in his subject's vision of red. He was most impressed with his result stating that 'this precise circumscription of the constitutive domain of a colour is a fact which seems to me new and worthy of being remarked.' [27, p178] But by this time it should hardly have been news considering Seebeck's work to which Wartmann referred several times. Wartmann was similarly impressed when his subject remarked that all the colours in the apparatus reddened upon a sudden change of illumination from overcast skylight to bright sun. Yet this observation is certainly similar to Seebeck's important point that colour discrimination depended upon luminance as well as gross perception of hues. The subject, who apparently belonged to Seebeck's class one, viewing colours in the apparatus under overcast conditions would have his colour sensitivity well reduced. But when the sun appeared he was able to see colour differences better with brighter light as expected from Seebeck's observation. Wartmann, however, simply suggested that some of the previously misnamed coloured glasses might have been correctly named had they been viewed under bright light but he made no attempt to repeat those observations. Wartmann let it go at that. Although he knew and praised Seebeck's work he did not really appreciate its details nor the subtleties of Seebeck's methods. Certainly he did not pursue his own investigations with anything like Seebeck's energy and the whole tenor of his work is inferior by comparison. Unfortunately for those relying on his review he never gave sufficient details of Seebeck's work so one could not really understand what Seebeck had accomplished. It is clear from Wartmann's own procedures, originally reported orally in 1840, that investigators of colour vision were still groping for reliable diagnostic methods while Seebeck's memoir went largely unappreciated.

One interesting point, however, Wartmann did make. In referring to Fraunhofer lines he stated 'it may further be interesting to remark, that D*** sees, in the whole extent of the solar spectrum which is luminous to him, the black stripes of Fraunhofer exactly as an ordinary eye . . . Daltonians therefore perceive darkness where it exists to us also.' [27, p180] Yet considering darkness as the absence of radiation, one could hardly expect otherwise. For the seriously colour defective the problem is really different; in the presence of light where the normal observer perceives an object with colour and good intensity, the colour defective may see it only as a shadow or darkness as was pointed out by Dalton himself. Even if seen with colour the object's brightness is reduced for the colour defective observer. Therefore, for a person of Seebeck's class two especially, Fraunhofer's lines in the red will be invisible since the red portion of the spectrum itself is invisible. This point was overlooked by Wartmann.

Probably Wartmann's most important contribution was an examination of all the various explanations of colour blindness. With one exception, we have already encountered them all and shall collect them here in order to show the range of choice possible at the time.

(*a*) Dougald Stewart (Scottish philosopher) called attention to differences of colour perception among men in 1792 in *Philosophy of the Human Mind*. He attributed it to a weakness in training or poor conception of colours rather than a defect of the eye or perceptual mechanism. Strangely enough, however, Stewart was himself colour blind, according to Brewster, having noted that he could not discriminate between the apples and the leaves of an apple tree except by their form and size.

(*b*) Dalton hypothesised a blue vitreous humour which by absorption prevented red light from reaching the retina.

(*c*) Goethe thought that colour blindness results from a lack of blue sensation, colour blind people seeing only a weak purple, rose-red and pure red. However, from Goethe's description of his two subjects they would lie in Seebeck's class one.

(*d*) Brewster (i) hypothesised that the eye is insensitive to colour at one extreme of the spectrum (proposed in analogy to work by Wollaston on limits of hearing).

(*e*) Brewster (ii): the retina, not the humours of the eye, is blue tinged and hence absorbs red light. (This followed (i) above and was based on the presumption that the choroid was the seat of vision, a point Wartmann overlooked.)

(*f*) Phrenologists: Wartmann reported their opinion as follows: 'According to the phrenologists, the distinction of colours is a faculty which does not depend upon the eye, but on a particular part of the brain, which they term the *organ of colours*. Those individuals in whom the portion of the brain which is immediately above the eye, beneath the eyebrow, is largely developed, possess in a high degree the faculty of distinguishing colours. It is thus the imperfection of this organ which is the origin of Daltonism.' [27, p184]

(*g*) George Wilhelm Muncke, Professor of Physics at Heidelberg stated that the optic nerve is concerned with two primary colours, red and blue, and their (presumed) complementaries, green and yellow. Activity of the primaries is initiated by the chemical power of light, that of the complementaries by the heating power of light. So the nerves are more or less reactant to the heating power of colours. Yellow is the best at heating the nerve which explains why it is seen best by colour blind people; red is the poorest and in colour blind people fails to heat sufficiently.

(*h*) Young and Herschel hypothesised that defective colour vision is due to inactivity or absence of one colour receptor out of a set of three.

For a science that didn't exist before the last decade of the eighteenth century, it is rather remarkable that all these possibilities existed shortly afterward. Wartmann gave the current objections to them as well, except in case (*g*). Stewart's hypothesis could not be reconciled with the facts. Dalton's hypothesis together with that of Brewster's (ii) were rejected on the basis of anatomical knowledge. (Note that Dalton was still living at this time.) Goethe's hypothesis was dismissed as not suiting the facts, while Brewster's (i) hypothesis likewise miscarried in terms of a bad analogy. Of the phrenologists he had no formal opinion but in a footnote he hinted 'I must say that among the Daltonians whom I have examined, the organ in question was in general but little evident, except in two of them, in whom it was remarkably prominent.' [27, p184] He even thought that colour blindness might in some cases be recognised by some external sign. Hence, without giving his blessing to phrenology, he did not dismiss it lightly either. Wartmann had no comment on Muncke's concoction which has not been heard of since. Finally Young's hypothesis was passed over with the comment that it is only a formula of facts although it is the only one of the hypotheses to which there is no objection.

Actually Young's hypothesis and Brewster's (i) are not dissimilar and conceivably could be reconciled with each other. Dalton's hypothesis and

Brewster's (ii) are also similar and contain an element of Young's thinking in that they prevent one colour from activating a visual sensation. Even Goethe recognised a lack of visual sensation to one colour; but his choice of blue as the one missing violates all the data known at the time. Colour blindness due to lack of blue sensation was not to be distinguished until long afterward and is, in fact, quite rare. Interestingly enough, there have been any number of hypotheses on the causes of defective colour vision since Wartmann's time and the subject is still very much under active research.

Whatever the merits of Wartmann's own investigations his exposition was an important one for several reasons. It was published in several European journals and in Britain. An oral report of his work given at the 1841 meeting of the British Association was published in the Report for that meeting and also in *Silliman's Journal* in America. Hence Wartmann's paper received wide attention as a result of which Wartmann was regarded as Europe's leading authority on 'Daltonism'. Small wonder then that Brewster, in defending his triple spectrum ten years later, appealed to Wartmann to defend his colour vision theory. Wartmann's paper was highly esteemed especially for its review of the literature and on this he was indeed meticulous, hardly missing any publication; yet much of this had previously been performed by Seebeck from whom Wartmann certainly took many of his references. Perhaps its main weakness was a lack of detail concerning Seebeck's work. As we have seen, Wartmann probably did not fully appreciate Seebeck's accomplishments. Nevertheless it was Wartmann who was read and much less so Seebeck, so ultimately it was he who brought colour blindness before the consciousness of other scientists. From Dalton to Wartmann, it had taken some 45 years for colour blindness to evolve from antiquarianism to a serious research field. The succeeding story will show that there was a direct link between Wartmann and Maxwell, whose work initiated modern methods of investigation.

One person who relied heavily on Wartmann's review was George Wilson, Regius Professor of Technology at the University of Edinburgh. Wilson collected a considerable body of statistics on colour blind people in 1852–3, publishing his results in several articles from November 1853 to December 1854. Unfortunately his reliance on Wartmann misled Wilson as to the true nature of Seebeck's work which he knew only through Wartmann's brief statements. Not surprisingly then Wilson mostly asked his subjects to name colours (from Hay's *Nomenclature of Colours*) or to describe their own confusions rather than employing Seebeck's method for eliciting detailed information. To a small extent however he did ask for matching or sorting of coloured wools. However, if Wilson can be faulted for this his work was otherwise of high calibre and was well received. Indeed the entire series of papers was reprinted in book form in 1855 receiving yet wider circulation. [33] Of special interest is a supplement to the book consisting of James Clerk Maxwell's first news of his own colour experiments. This was in the form of a letter to Wilson especially solicited by him for inclusion in his book and will be treated in Chapter 8. In general, Wilson's book brought the subject to the attention of the general public, as opposed to scientists only, confirming that colour blindness was actually quite prevalent in the population.

An interesting aspect of the work was its attempt to draw some practical conclusions. A first illumination in this direction had been given in a letter to the *Athenaeum* in January 1853. The writer stated that railway signals which employed red, green and white lamps at night may be confused and hence lead to accidents. He proposed that the red light for danger and the green light for caution could under certain circumstances be viewed together and mistaken for the white safety signal. Hence he concluded:

> It is not improbable that some of the accidents which have occurred in railway travelling have arisen from the colours of the lights shown being indistinctly seen; perhaps, from a confusion of rays from two or more lamps. In some cases most contradictory evidence has been given as to the colour of the signal shown. [34]

Although there was no hint of colour blindness as the cause of accidents, his letter was read by Wilson, then conducting his experiments, who replied to the *Athenaeum* in April, 1853. [35] He pointed out that the problem of identifying coloured railway signals was more important in daytime when coloured flags and coloured forms were used and that misreading of these signals could well result from colour blindness on the part of the so-called railway servants. [35] The red and green signals were exactly those confused by colour blind people, Wilson stated, and people who could not distinguish these colours should certainly not be employed as signalmen or engine drivers. He recommended that all such people should be tested for so dangerous an incapacity as that of colour blindness. But the most important part of his letter was its final paragraph where he asked the *Athenaeum* readers to contribute to the statistics of colour blindness by sending him the details of their difficulties. His replies were numerous and together with his own voluminous investigations led to the first good collection of statistics on colour blindness.

Wilson's 17 case histories, both those communicated to him and those investigated by himself, showed nothing particularly new. Still labouring under the communication problem so long ago pointed out by Seebeck, he was unable to classify his subjects well although he did recognise that several were truly blind to red. However, he attacked the statistical aspects of colour blindness on a grand scale. An indication of the magnitude and scope of his investigations can be seen from Table 7.2.

Since ordinary soldiers came from working class backgrounds while most students were of middle to upper class origins the data show that not only is colour blindness quite prevalent but that it knows no social boundaries. (Considering the scope of the attempt, Seebeck's method for these purposes would have been inappropriate.) It was well known by then that colour blindness was not rare, but exactly how prevalent it was remained unknown. Wilson showed that 5.6% of the (male) population must be affected, a rather good first statistical result. It is now known that the figure (all categories

Table 7.2. Extent of colour blindness, as ascertained by the examination of 1154 persons at Edinburgh, in 1852–53.

Profession	Number of persons examined	Confounded red with green	Confounded brown with green	Confounded blue with green	Number of persons colour blind	% of persons colour blind	Proportion of persons colour blind referred to unity
4th Infantry	437	6	13	12	31	7.1	1 in 14.1
7th Hussars	177	5	2	6	13	7.3	1 in 13.6
Artillery soldiers, Leith Fort	123	2	1	2	5	4.0	1 in 24.6
Professor Kelland's pupils, Edinburgh University	150	3	0	0	3	2	1 in 50
Edinburgh Police	158	1	2	2	5	3.1	1 in 31.6
Students of the Edinburgh Veterinary College	47	0	0	1	1	2.1	1 in 47
Medical and other attendants, Royal Asylum, Morningside	42	2	1	2	5	11.9	1 in 8.4
Dr G Wilson's pupils	20	2	0	0	2	10.0	1 in 10
	1154	21	19	25	65	Average 5.6	Average 1 in 17.7

Ratio of special colour blindness, as deduced from the table.

Confound red with green, 1.8%, or 1 in 55; confound brown with green, 1.6%, or 1 in 60; confound blue with green, 2.2%, or 1 in 46.

included) is about 7%. Most interesting is the rather high percentage of his students who were colour blind for it was their difficulties in Wilson's chemistry classes that originally stimulated him to this research. His fourth column is also interesting in hinting at people with difficulty at the blue end of the spectrum although the percentage is far too high and his methods too unreliable to give this any validity. Yet even the suspicion of such a possibility escaped Wilson which is surprising since he embraced Young's theory of colour perception and its colour blindness corollory of all the extant possibilities. Let us consider them briefly once more as now viewed in 1855, fifteen years after Wartmann.

Remarkably the phrenological position was still being discussed if only to finally refute it. That both Wartmann and Wilson considered it at all indicates that it could not be dismissed out of hand. It had at least a 'superficial logic!' Basically Wilson found that it did not fit the facts. Moreover, comparing the head of Dalton who was definitely red blind with that of Newton who was assuredly not colour blind, Wilson said,

> Those who have seen him, or who are familiar with his busts or portraits, will be aware that his head, which remarkably resembled that of Sir Isaac Newton, was like it in exhibiting prominent eyebrows. The superciliarly ridge in truth, projected very markedly so as to overhang the eyes. If any man would have been selected by phrenological signs as the possessor of an exquisite sense of colour, Dalton would have been. Yet assuredly, if Newton and Dalton had been contemporaries, and had simultaneously published and compared their descriptions of the colours of the rainbow, they must have been hopelessly perplexed to account for statements, so strangely conflicting, regarding a common object of vision. [33, p107]

This sufficiently showed the fallacy of phrenology: like the travellers in Brewster's fable, Dalton and Newton would have needed a third party to settle their colour differences. More to the point was that colour perception apparently depended on vision of three primaries, red, yellow (still) and blue. Since most colour blind people could see yellow and blue, at most only one-third of the 'organ of vision' should have been missing. Therefore one could still expect some prominence of the superciliary ridge even in colour blind people. Thus the whole idea was full of objections, yet in the end he could not bring himself to dismiss it altogether, for in his final remark he stated 'it is assuredly desirable that observations should be multiplied on this point, and that the condition of the brain in the colour blind should be carefully noted.' [33, p108] So ultimately he acted the part of the cautious scientist. Yet from the phrenologists' hypothesis, however, one might predict that our simian cousins must have a remarkable colour sense. How remarkable it is remains a matter of opinion but in this century colour vision has in fact been demonstrated in anthropoids and some other animal species. But whether colour blind individuals exist, as might be expected, with colour confusions remains to be seen. Certainly there will be a communications problem there!

The idea about colouring matter in the humours of the eye, the choroid or the retina was also rejected, based by now on numerous autopsies and especially recently by observation of the living eye with the ophthalmoscope, invented by Helmholtz. There was simply no evidence for it.

The whole review came down to the theory chosen by Wartmann as having no objection to it. The early idea of Young considering the absence or paralysis of one of the colour receptors in the retina was finally embraced. This could explain the absence of sensitivity to one colour and the colour confusion in the colour blind on the basis of abnormal mixing by the remaining two receptors. This theory had been adopted by Professor Philip Kelland, one of Britain's leading mathematicians and editor of Young's works and, perhaps significantly, one of Maxwell's teachers. Kelland argued persuasively in favour of a theory such as Young's at a meeting of the Royal Scottish Society of Arts in April 1853 and at Wilson's request wrote out his ideas. These were printed verbatim by Wilson who, in essence, accepted them completely.

The remainder of Wilson's review of colour blindness involved a discussion of possible cures for which he found none, and an application of the knowledge to railway signals and ship's navigational lights. In this he was undoubtedly the first to recognise that some accidents, especially those involving railways, could be accounted for by deficient colour vision. As noted before, not only were red and green signals exactly the colours confused by the colour blind, but even worse the coloured flags employed were often so dirty that both colours appeared rather dark. Even normal observers could have difficulty discriminating the colours especially at high speeds; for the colour blind it would be impossible. And of course the signalman himself could well set the wrong colours. Summarising these points Wilson said

> Our railway directors have, indeed, been most unfortunate in their selection of red and green for coloured signals, especially as seen through the day on flags. These colours are mistaken for each other by certain of the colour blind even when fresh and bright. [33, p125]

In consequence of this he again recommended colour testing of railway employees which was immediately instituted by the Great Northern Railway in Britain. In general the heavy colour dependency of signals at the time together with their other insufficiencies led to the introduction of new signal methods including semaphores whose message depended more on form and position than on colour.

We cannot leave this history without noting the origins of the term 'colour blindness.' Part of the recognition of this visual defect involves a search for an acceptable nomenclature. We have seen that Goethe suggested (incorrectly) the name 'acyanoblepsia' meaning a non-perception of blue, while 'Daltonism' was formally due to Prévost in 1827. That term, however, had been used in lectures in Geneva as early as about 1800. The eminent Czech physiologist J E Purkinje conceived of four categories of abnormal colour vision which he named *achromatopsis, chromatodysopsis, acyanoblepsia* and *anerythroblepsia.* The first refers to a total lack of colour vision, persons so affected seeing only black, white and grey. Purkinje was referring to cases like that of the *Philosophical Transactions* of 1684 and similar such medical cases and possibly to Harris whose colour vision was certainly very restricted. Seebeck in fact thought that Harris might just fit this category although he noted that reports of this kind mostly stemmed from incomplete observation. [32, p228] However, in terms of heredity (as opposed to mental disorder or injury induced causes) it is now known that this form of colour blindness comprises only about 0.003% of the population and almost certainly none of our writers observed such a case. Chromatodysopsis referred to abnormal perception of colour related mainly to brightness while the other two terms referred to missing visual sensations. The physiologist Szokalski in 1842 introduced 'chromato-pseudopsis' meaning a false vision of colours, a term adopted by Wilson. Finally Wartmann used the terms 'polychromatic Daltonism' and 'dichromatic Daltonism' to refer to Seebeck's classes one and two respectively. None of these terms, however, found much favour being considered either too awkward or not sufficiently suggestive of the phenomena. Compared to these, only 'Daltonism' seemed sufficiently easy to use and sufficiently flexible to provide both a noun and an adjective. This, of course, was the popular term used on the Continent.

Naturally enough British investigators objected to perpetuating Dalton's name this way. Their view was expressed by William Whewell speaking in commentary to Wartmann's report at the British Association meeting in 1841. Whewell 'doubted the propriety of the name given to this defect; few persons would desire to be immortalised through the medium of their defects, and Dalton, least of all, required such a means of handing down his name to posterity.' [36] Whewell's own suggestion was probably worst of all, being 'idioptcy' and 'idiopts' for adjective and noun respectively. This unfortunate terminology was rejected out of hand by everyone. In a commentary on Wartmann's paper in 1844 Brewster also objected to Wartmann's continued use of the 'offensive name of Daltonism.' [37, p139] Briefly noting all the various terms then available together with their difficulties, Brewster stated 'we have used the word *colour blindness* (sic) because it indicates simply blindness to one or more colours.' [37, p140] Indeed he had used it previously in this commentary but now brought the term to explicit attention. Brewster's term found ready acceptance. In the English translation of Wartmann's *Memoir on Daltonism* (appearing in 1846) the translator added to the title in parentheses 'or colour blindness' apologising in a footnote in giving a faithful translation for continuing Wartmann's use of 'Daltonism'. Wilson also adopted Brewster's term which he used interchangeably with 'chromato-pseudopsis' and Maxwell similarly adopted it in his first two papers on colour in 1855, to be discussed presently. Probably the most important change in attitude came with its acceptance by

Helmholtz (Farbenblindheit) in his *Handbuch der Physiologischen Optik*, published in 1867. Actually Part II of that work where the term is used was first published in 1860, so Brewster's term by then had definitely entered the literature.

Unfortunately even this term is not entirely accurate implying exactly as Brewster said a real blindness to one or more colours. But Seebeck had noted a whole range of abnormal colour vision from severely altered perception to nearly normal. The term would hardly be appropriate to many of the approximately normal. Moreover 'colour blindness' implies perceptual difficulty only with hue discrimination while Seebeck had also demonstrated confusions involving luminance. In fact errors may also involve saturation and those with abnormal colour perception may experience difficulties with hue, luminance and saturation separately or in combination. So the term 'colour blindness' is still rather inadequate in this respect. A term now much used is 'anomalous colour vision'. Nevertheless in 1844 Brewster's term was reasonably justified since the subtlety of confusions due to brightness was still unappreciated and that of saturation was unknown. For British investigators, in any case, it was far superior to 'Daltonism'. Of all the competing terms Brewster's was the most suggestive of the difficulties involved and the easiest to employ. For these reasons it has entered the language as a popular term denoting abnormal or anomalous colour vision.

NOTES

1 The subject is not mentioned, for example, in any of the following works where it would be expected to appear: Joseph Priestley 1772 *The History and Present State of Discoveries Relating to Vision, Light and Colours* (London); Ronchi, Vasco 1939 *The Nature of Light: An Historical Survey* (English translation Baracas, V 1970 (London: Heinemann Educational Books)); Lindberg, David C 1976 *Theories of Vision from Al-Kindi to Kepler* (Chicago: University of Chicago Press); Halbertsma, K J A 1949 *A History of the Theory of Colour* (Amsterdam: Sivets and Seitlinger); Sarton, George 1927–48 *Introduction to the History of Science* (Washington: Carnegie Institution of Washington; and Baltimore: Williams and Wilkins)

2 de la Hire, Philippe 1680 Dissertation sur les differences accidens de la vue *Mém. Acad. R. Sci. Paris* **IX** 539

3 Boyle, Robert 1688 Some uncommon observations about vitiated sight, appended to *A Disquisition about the Final Causes of Natural Things* (London). Reprinted 1772 in *Works* ed Thomas Birch vol V (London) 445–52

4 Priestley, Joseph 1772 *The History and Present State of Discoveries Relating to Vision, Light and Colours* (London)

5 1684 *Phil. Trans. R. Soc.* 736

6 Huddart, John 1777 *Phil. Trans. R. Soc.* **67**

7 1778 *Phil. Trans. R. Soc.* **68** 611

8 Lichtenberg, J C and Voight, J H 1781 *Gothaisches Magazin fur das Neuste aus der Physik und Naturgeschichte* 57–81 as mentioned in Walls, Gordon L 1956 The G Palmer story *J. Hist. Med. Allied Sci.* **XI** 90–2

9 *Encyclopaedia Britannica* 1797 3rd edn (Edinburgh). Harris's case is in volume

5 under the heading 'Colour'. Scott's case is mentioned in a single sentence without his name but with the correct citation under the heading 'Sight'.
10 Stewart's colour vision is described in Brewster, David 1853 *Treatise on Optics* (London) p439
11 Dalton, John 1798 Extraordinary facts relating to the vision of colours; with observations *Mem. Lit. Phil. Soc. Manchester* 5 This was read to the society on 31 October 1794.
12 Rosco, Henry E 1895 *John Dalton and the Rise of Modern Chemistry* (London) p70
13 Wild, C R 1848 *A History of the Royal Society* (London) as quoted by Hall, M B 1975 Science in the Early Royal Society in *The Emergence of Science in Western Europe* ed Maurice Crosland (London: MacMillan). References are to the 1975 version.
14 Letter from Dalton to Joseph Dickinson, as quoted by Rosco [12]
15 Brewster, David 1833 *Letters on Natural Magic addressed to Sir Walter Scott* (London) pp32–3
16 Young, Thomas 1807 *A Course of Lectures on Natural Philosophy and the Mechanical Arts* vol II pp315–16. The remark is made in his *Catalogue of Works Relating to Natural Philosophy* as an annotation to the entry of Dalton's paper on his colour blindness.
17 Letter from Sir J F W Herschel to Dalton, May 20 1833 [18, pp25–6] The emphasis is Herschel's.
18 Henry, William Charles 1854 *Memoirs of the Life and Scientific Researches of John Dalton* (London: Cavendish Society)
19 Wright, W D 1968 The Unsolved problem of Daltonism, in Cardwell, D S L (ed) *John Dalton and the Progress of Science* (Manchester: Manchester University Press) p312
20 Wardrop, John 1808 *The Morbid Anatomy of the Eye* 1st edn (London)
21 von Goethe, Johann Wolfgang 1810 *Theory of Colours*. (English translation 1840 by Charles Eastlake). Reprinted 1970 (Cambridge, MA: MIT Press) p47
22 Butter, John 1822 Remarks on the insensibility of the eye to certain colours *Edinb. Phil. J.* **6** (The paper was also published in 1882 *Trans. Phrenol. Soc.*)
23 Nicholl, Whitlock 1822 *Trans. Medico-Chirurigal Soc. London* **7** 447
24 Nicholl, Whitlock 1822 Remarks on a peculiar imperfection of vision with regard to colours *Ann. Sci.* New Series **III** 128–131
25 Harvey, George 1826 On an anomalous case of vision with regard to colour *Trans. R. Soc. Edinb.* **10** 253–62
26 Prévost, Pierre 1827 Sur la vision extraordinaire de J B *Bibliothèque Universelle des Sciences, Belles Lettres, et Arts* **35**
27 Wartmann, Élie 1843 Memoir on Daltonism *Memoires de la Société de Physique et d'histoire Naturelle de Genève* English translation 1846 in *Taylor's Scientific Memoirs* (London) p158
28 Brewster, David 1826 On the invisibility of certain colours to certain eyes *Edinb. J. Sci.* **86**
29 Brewster, David 1826 Account of two remarkable cases of insensibility in the eye to particular colours *Edinb. J. Sci.* 153–9
30 Herschel, John F W 1845 Article on light *Encycl. Metrop.* **III** 432–5
31 Colquhoun, Hugh 1829 An account of two cases of insensibility of the eye to certain of the rays of colour *Glasgow Med. J.* **2** 12–21
32 Seebeck, A 1837 Uber den bei mancher Personen vorkommenden Mangel an Farbensinn *Ann. Phys., Lpz.* **42**
33 Wilson, George 1855 *Researches on Colour Blindness* (Edinburgh)
34 Tyndall, W H 1853 *Athenaeum* January 29 p139

35 Wilson, George 1853 *Athenaeum* April 2 p418

36 Reported in *Athenaeum* 1841 p669 in its description of the meeting of the British Association for August 1841

37 Brewster, Sir David 1844 Observations on colour blindness, or insensibility to the impressions of certain colours *Phil. Mag.* Ser.3 **25**

8
James Clerk Maxwell: The Beginning of Modern Methods

We have seen so far that two critical events in the theory of colours occurred in 1852 and 1853. Helmholtz had shown that colour mixtures of light and pigments obey different laws so that neither method can give information about the other. He had also elucidated the physical nature of pigment mixtures while providing a reliable method in his V-slit experiment to make colour mixtures of light.

The second event was Grassmann's paper of 1853 which provided a theoretical framework within which Helmholtz's work could be understood. The mathematical variables of colour were for the first time clearly enunciated, the possibility of colour equations was demonstrated and the vector nature of the equations was shown. In addition, one could now see that Newton's colour circle, Mayer's colour triangle and Mayer's and Lambert's colour pyramids were simply different aspects, or different graphs, of the three variables of colour. Moreover, Grassmann prompted Helmholtz to extend his observations of spectral complementary colours to at least seven pairs. Therefore, by 1853 there was a good qualitative understanding of colour mixtures of light together with a quantitative but as yet unconfirmed theory of the process.

In a real sense these successes formed a new paradigm for further work in colour vision. Besides completely reformulating concepts of colour and colour vision its success lay in the fresh questions it raised. Thus one could now ask: (i) within what limits do these concepts of colour apply to normal eyes, (ii) specifically how do colour blind eyes differ from normal eyes, and (iii) can this theory of colour vision be brought within the scope of the wave theory of light?

The first question is interesting in terms of an age-old question that everyone has asked himself. What does it feel like to be someone else? With the theoretical possibility of producing quantitative colour data based on Grassmann's theory it became possible to give an answer to this question within the narrow context of colour perception. We shall see in this chapter that the great physicist James Clerk Maxwell produced experimental

James Clerk Maxwell
(courtesy of Burndy Library).

methods which at once gave quantitative limits within which normal eyes are the same and also answered the second question in terms of Young's hypothesis. Mathematical attempts at answering the third question were essentially unsuccessful but will be noted in Chapter 9.

MAXWELL'S BACKGROUND

Before beginning this discussion, however, it is interesting to note the factors which influenced Maxwell's choice of colour vision as a research area. Only rarely is it possible to determine why someone has chosen a particular field of interest; indeed it is probably difficult to determine the reasons for oneself. Certainly for a person of Maxwell's far ranging genius any problem in physics was grist for his mind as were several purely mathematical questions. Hence, strictly speaking we cannot really tell why Maxwell was interested in colour vision. Nevertheless we can trace in Maxwell's life a thread of intrinsic interest together with certain important later influences on him.

Born in 1831, Maxwell exhibited early an extraordinary eye for colour matching and colour harmony. Before the age of eight, for example, he was already imitating his mother's and aunt's knitting with original patterns of harmonised colours. [1, p36] Of his choice of clothing, his biographer, Lewis Campbell, Maxwell's school companion and life-long friend, states 'an "esthetic" taste might have perceived in its sober hues the effect of his marvellous eye for harmony in colour.' [1, p162] Campbell also noted that in many long talks about poets and general literary subjects Maxwell 'in his critical studies of modern poets . . . used to note their fondness for particular colours, e.g., the uses of "white," "red," "black," "ruby," "emerald," "sapphire," in Tennyson and Browning' [1, p163] Perhaps most significant was a scientific toy which came into vogue in the 1830s known variously as the 'stroboscope' or 'magic disc'. It consisted essentially of a disc revolving before the eye, the observer viewing figures through slits cut at equal intervals near the rim of the disc. Then, as in modern cinematic animation, the figures placed in graduated positions gave an illusion of motion. For this instrument the young Maxwell devised many ingenious animations including several coloured patterns whose designs opened and closed. [1, p38] Unwittingly Maxwell performed spinning disc experiments with colour at an early age. There is no chronological connection between these childhood amusements and his later colour top experiments, the subject of this chapter; yet Maxwell possessed an exceptional memory. Coupled with later influences, these early experiences were probably not lost upon him. In fact his early sensitivity to colour and facility in design remained with him life long. Of this William Garnett, collaborator with Campbell on Maxwell's biography, who knew Maxwell in his last years, says that he 'had great facility for designing, and would frequently amuse himself by making curious patterns for wool-work. His designs are remarkable for the harmony of the colouring.' [1, p487]

It does not follow, of course, that intrinsic interest blossoms into professional interest regardless of the ability of the man, but other influences of his education, both formal and informal, almost certainly brought Maxwell to a professional interest in colour. At the age of fifteen, Maxwell's father began taking him to meetings of the Royal Society of Edinburgh and the Edinburgh Royal Society of Arts. [1, p73] Scotland at the time was a world centre for scientific research, being the home of such persons as William Thomson, David Brewster, J D Forbes, and many others of lesser renown today but then eminent persons. Among them were D R Hay, George Wilson and Philip Kelland, the latter being one of Britain's leading mathematicians. Thus papers read at the Royal Society of Edinburgh were often of outstanding importance in an era when physics was reaching a new maturity. For example, William Thomson's classic papers on thermodynamics were read before the Society and published in its *Transactions*. Significantly, Brewster, Hay, Forbes and Wilson, all interested in aspects of colour, all resided in Edinburgh where Maxwell went to school as a boy. Within this milieu Maxwell spent his teenage and perhaps most impressionable years meeting some of the most important scientific personalities of the century; here he witnessed the announcement of important discoveries. An amusing anecdote in this respect is related by Maxwell's older contemporary, William Swan, in a letter to his biographers. At the 1850 meeting of the British Association in Edinburgh, Maxwell, then just nineteen years old rose to dispute a point with the great Sir David Brewster concerning the phenomenon of vision known as 'Haidinger's Brushes'. The astonished dignitaries looked on with an air that was 'half-puzzled, half-anxious, and perhaps somewhat incredulous' [1, p489, p144] while this beardless youth disputed with one of the accepted masters. Maxwell, however, overcame his shyness, said all he meant to say and gained the respect of his listeners. Unfortunately what Maxwell said is unrecorded as is Brewster's reply, if any. Brewster, however, had attributed the effect to polarisation of light in the cornea and crystalline lens, with which Maxwell disagreed. Six years later he showed that the effect derived from polarisation in the retina, not in the crystalline lens. [2]

In meeting the leading members of his future profession, probably the most influential for our purposes were D R Hay and J D Forbes. Hay, for example, had just brought considerable scientific attention to himself by his book *First Principles of Symmetrical Beauty* in which he attempted to apply mathematical principles to the subjects of form and colour. It was this subject which led Maxwell at the age of 15 to his first original scientific work on the construction of ovals. Luckily his father knew Forbes to whom he sent the manuscript, who in turn sent it to Kelland and following revisions it was read (by someone else) to the Royal Society of Edinburgh. It was not thought proper that a young boy should address the Society. This early acquaintanceship with Forbes continued upon Maxwell's entry at the age of 16 into the University of Edinburgh where Forbes was Professor of Natural Philosophy. Here Maxwell was Forbes' student and the acquaintanceship soon ripened into real friendship. The historian J T Merz has said that within the Scottish universities there was a congenial atmosphere and spirit which allowed and fostered the possibility of intimate acquaintanceship between the professors and their students. [3] Moreover, until the mid-nineteenth century most of the great English men of science worked outside the university system (of this, among others, Davy, Wollaston, Young, Dalton, Faraday and Joule are notable examples) while most of the eminent Scottish men of science were university professors, examples being such people as Black, Playfair, Thomson, Leslie and Forbes. Thus not only was there a congenial atmosphere within the Scottish universities but it was almost certain that a nascent scientist would meet there active scientists who would later become his colleagues. Within this environment Maxwell became a particular favourite of Forbes who granted him free access to class apparatus for his own original experiments. [4]

Forbes' own interests at the time included that of colour, so it is not surprising to learn that his favourite pupil was present during his colour experiments. As Maxwell himself said, 'the experiments of Professor J D Forbes, which I witnessed in 1849, first encouraged me to think that the laws of this kind of mixture might be discovered by special experiments.' [5, p145] [6] It was then that Forbes had attempted to form colour equations and had made the surprising discovery that yellow and blue yielded grey rather than green on the rotating disc. To determine whether similar results could be obtained without motion Forbes had proposed other experiments but as previously noted he did not pursue this work. Maxwell, however, did do so, confirming for himself in August 1852 that mixtures of spectral colours gave results consistent with the spinning disc. [5, p144] [7] In fact his approach was similar in spirit to Helmholtz's V-slit method and he naturally arrived at similar conclusions although Helmholtz's paper was then unknown to him. Both men performed essentially equivalent experiments at the same time, an example of simultaneous discovery. However, these early experiments of Maxwell went unpublished as they did not constitute a formal research programme for him as they did for Helmholtz. Helmholtz, already an established research scientist, had set out specifically to overthrow Brewster's triple spectrum and to construct a new concept of colour. Maxwell, by now a busy undergraduate at Cambridge, almost certainly viewed himself as simply pursuing his basic interests in optics and vision and continuing his former teacher's experiments during the long summer holiday. For him it constituted important experience for his major experiments yet to come. In any case it seems that Maxwell's intrinsic interest in colour received a professional stimulus from his association with Forbes and no less so with Hay and from the ongoing discussions on colour in Edinburgh. From Forbes especially ideas concerning possible quantitative colour data had been planted in Maxwell at an early age, certainly before October 1850 when he left Edinburgh to go up to Cambridge, so Helmholtz's papers of 1852 and 1853, and Grassmann's work of 1853 fell on fertile ground in the person of Maxwell. By early March 1854 in his last year at Cambridge, Maxwell had already performed some preliminary observations convincing himself that his proposed experimental method, the colour top, yielded reliable data. Looking to the near future in this regard, Maxwell wrote to William Thomson saying,

> the subject of colour as perceived by the eye is one I intend to investigate more fully. I have satisfied myself that by using the aid of others in every experiment, and shutting out obvious sources of error, very exact results may be obtained about the equivalence of two colours as regards the eye. [8]

These remarks refer to his colour top method of obtaining colour vision data which forms the substance of this chapter. The letter even embodies the title of Maxwell's first major publication on colour, *Experiments on Colour as perceived by the Eye, with remarks on Colour Blindness.* [5] Upon attaining his degree in the Spring of 1854 and then having time for his own researches, Maxwell's way was opened to a major contribution towards understanding colour vision. By the end of November 1854 he had been 'spinning colours a great deal' [9] and was ready with data and conclusions which he reported briefly in a letter to George Wilson at Wilson's request. There in a concise summary of chronology, but not necessarily influence, Maxwell stated:

> I have put down many things simply to indicate a way of thinking about colours which belongs to this theory of triple sensation. We are indebted to Newton for the original design; to Young for the suggestion of the means of working it out; to Professor Forbes for a scientific history of its application to practice; to Helmholtz for a rigorous examination of the facts on which it rests; and to Prof Grassmann . . . for an admirable theoretical exposition of the subject. [10, pp158–9]

MAXWELL'S EXPERIMENTAL METHOD

Colour combinations by a revolving disc or top had been shown by Helmholtz in published work and by Maxwell privately to give results consistent with combinations of spectral colours. Therefore this technique was raised from a laboratory ambiguity to a legitimate research technique. It was an easy method to employ and was chosen by Maxwell for his first investigations. Maxwell's top, however, differed from those of his predecessors such as Young and Helmholtz in a very important manner. They had painted the disc thereby making the colours non-adjustable in area. But Maxwell, following Forbes' method, used coloured papers, designed with slots so they could easily fit over the axis of the disc [5, Plate I] (see Figure 8.1). These papers, at first prepared by Maxwell's old acquaintance, Hay, could then lie flat on the disc and easily slide over each other. Therefore one could place any number of coloured papers on the top and easily adjust the areas exposed. Upon screwing down a handle above them, the top was set in a holder and spun by pulling a string [5, Plate I] (see Figure 8.2). Maxwell described the procedure as follows:

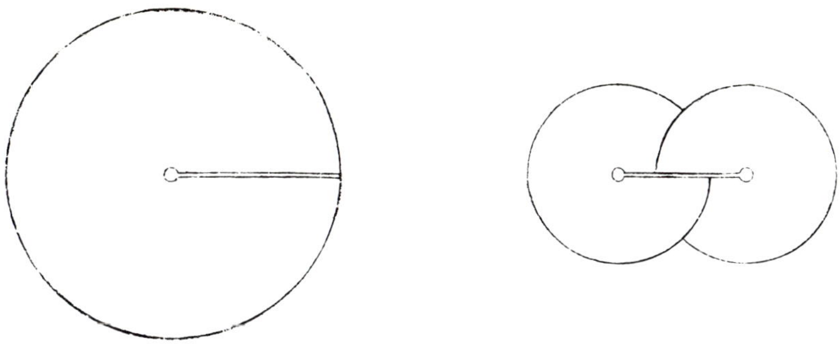

Figure 8.1 Coloured papers with slots to fit on the colour top.

> the form in which the experiment is most manageable is that of the common top. An axis, of which the lower extremity is conical, carries a circular plate, which serves as a support for the discs of coloured paper. The circumference of this plate is divided into 100 equal parts, for the purpose of ascertaining the proportions of the different colours which form the combination. When the discs have been properly arranged, the upper part of the axis is screwed down, so as to prevent any alteration in the proportions of the colours. [5, p126]

We may note immediately the use of a scale on the perimeter divided into 100 parts, as had been Forbes' disc. The disc containing several papers is shown in Figure 8.3. [5, Plate I] The perimeter of the circle was simply marked off by the black triangles (▲) representing 10 divisions, and the dots (:) for 5 divisions. Therefore, reading the scale consisted of counting with these aids the number of divisions exposed by each sector of paper.

Although the adjustable colour sectors represented an important innovation, there was yet one more even more important. A second, smaller set of coloured papers was placed on top of the first. These were made half the

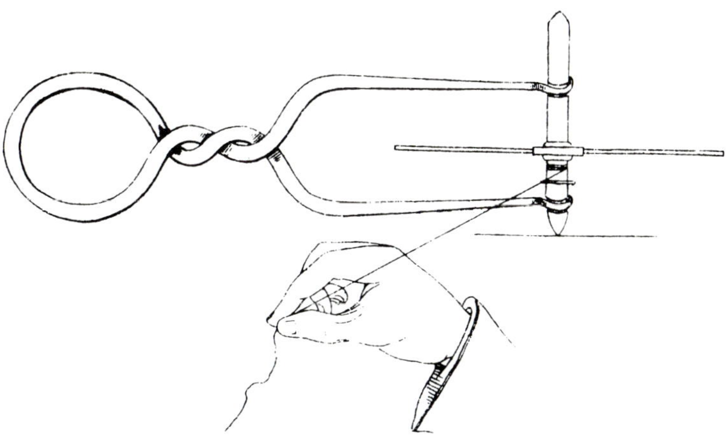

Figure 8.2 Maxwell's colour top.

diameter of the lower papers [10, p157] leaving the outer portion of the larger discs exposed. As seen in Figure 8.3 the two smaller papers are black and white. With this arrangement one saw upon spinning the disc two colour combinations in such a way that 'the resultant tint (hue) of the first combination will then appear in a ring round that of the second, and may be carefully compared with it.' [5, p126] (Maxwell is holding his top in the portrait on p154.)

(*a*) (*b*)

Figure 8.3 Coloured papers on the top. (*a*) Schematic; (*b*) actual (from Gunther, R T 1937 *Early Science in Cambridge*, courtesy of Oxford University Press).

The latter phrase 'may be carefully compared with it' is most significant. Grassmann had shown just two years earlier that equations of colour were possible in principle and that experimentally such equations would take the form of photometric colour matches. This was the central idea of his argument described in Chapter 6. But comparison of different lights in a photometer requires the two illuminated fields to adjoin. Forbes in his attempts to obtain colour equations had not satisfied this important experimental criterion. He had simply spun several discs on a machine simultaneously using colours on one and black and white on another. [11] Hence Maxwell provided here the last step necessary to obtain colour equations as required by Grassmann. His colour top actually executed Grassmann's thought experiment (pp 95–6) in which colours were to be matched by lights projected onto adjoining screens. The question then remaining was: What colours should be chosen to affect the colour matches? In answering this question it is interesting to briefly glimpse Maxwell's experimental philosophy.

The success of physical investigations, Maxwell said in later years, 'depends on the judicious selection of what is to be observed.' [12] In the case of colour investigations this was critically important because of the subjective nature of the sensations themselves. On this Maxwell said,

> Now, although every one (sic) experiences these sensations, and though they are the foundation of all the phenomena of sight, yet, on account of their absolute simplicity, they are incapable of analysis, and can never become in themselves objects of thought. If we attempt to discover them, we must do so by artificial means; and our reasonings on them must be guided by some theory. [10, p154]

The theory to be adopted, of course, was essentially that of Grassmann and Young within which lay the formal explicit conception of three variables of vision. After giving a synopsis of existing theory regarding the number of colour sensations Maxwell went on to say:

> the theory which I adopt assumes the existence of three elementary sensations, by the combination of which all the actual sensations of colour are produced. It will be shown that it is not necessary to specify any given colours as typical of these sensations. Young has called them red, green, and violet; but any other three colours might have been chosen, provided that white resulted from their combination in proper proportions. [10, p155]

Thus the particular choice of colours was not really a crucial question as long as it satisfied the latter criterion. Thanks to Helmholtz, one was not restricted in experiments to a particular set of 'primary' colours. Nevertheless Maxwell chose red, green and blue papers to play a role similar to that of primary colours.

Maxwell felt called upon to explain his choice of green rather than yellow and undoubtedly did so in an effort to avoid criticism. For although the observational work of Helmholtz and Grassmann's theory undoubtedly represented a small scientific revolution, it was a revolution, as all such conceptual changes are, for only a narrow part of the scientific community. For example, such leading personalities as Wilson and Kelland, actual contributors to the field of colour perception, were still referring to yellow as a primary colour in 1855, three years after Helmholtz's work and two years after its appearance in English in the *Philosophical Magazine*. And they did so despite Helmholtz's results after Grassmann having been announced first in Britain at the 1853 meeting of the British Association. Brewster, of course, tenaciously maintained his position on yellow as a primary spectral colour to the end. Thus it remained to convince the larger community that a real change in viewpoint had emerged. In this respect it has been said that:

> the decision to reject one paradigm is always simultaneously the decision to accept another, and the judgement leading to that decision involves the comparison of both paradigms with nature *and* with each other. [13, p77]

For Maxwell this process had already taken place in respect to the choice of green rather than yellow. But for the larger scientific audience at which his work was aimed, it is questionable how many had yet accepted or even knew of this new view. The old concept of primaries was still very much alive. This was confirmed by Maxwell himself with wry humour in a letter to his father dated 5 May 1855. Describing how he had demonstrated his colour top before the Cambridge Philosophical Society, Maxwell said 'there were sundry men who thought that Blue and Yellow make Green, so I had to undeceive them.' [1, p212] Even the great Sir George Gabriel Stokes remained to be convinced allowing that Brewster's experiments if not his triple spectrum were correct. [14] Therefore, it was no simple matter to suddenly publish a work in which yellow was replaced by green as a primary colour. In other words, besides the purely logical procedure mentioned above of comparing theory with nature, a process of argumentative persuasion was necessary as well to make an end to the old paradigm of yellow rather than green. [15]

In this respect, then, Maxwell said early in his first detailed description of his experiments, that his red, green and blue papers did not really represent primary colours. [5, p127] Instead they were 'simply specimens of different kinds of paint, and the choice of these was determined solely by the power of forming the requisite variety of combinations.' [5, p127] With the first part of this statement Maxwell introduced the red, green and blue primaries, as he surely viewed them, in the context of the older theory thereby making it more palatable and more easily acceptable. Cleverly employing the old familiar idea of paints, he introduced a major new idea in a way that seemed to be just a variation on the old. The second part of the statement is a powerful experimental argument, short as it may seem, for he continued:

> if red, blue and yellow had been adopted, there would have been a difficulty in forming green by any compound of blue and yellow, while the yellow formed by vermilion and emerald green is tolerably distinct. [5, p127]

The meaning of this was really that red and green *light* mixtures form yellow, while blue and yellow light mixtures do not form green. This is not the case with pigment mixtures, of course, where blue and yellow do give green. Hence, had he chosen yellow as a primary colour he could not have produced green from any combination of his three primaries and the number of colour combinations possible would have been much reduced. (This is discussed in detail in Chapter 9.) Knowing this in advance, the obvious choice therefore was to use green and not yellow. Maxwell's choice of colours, shown in Table 8.1, [5, p127] was not limited to one selection of each. Within each group

Table 8.1. Maxwell's choice of coloured papers for the top

Vermilion	V	Ultramarine	U	Emerald Green	EG
Carmine	C	Prussian Blue	PB	Brunswick Green	BG
Red Lead	RL	Verditer Blue	VB	Mixture of Ultra-marine and Chrome	UC
Orange Orpiment	OO				
Orange Chrome	OC				
Chrome Yellow	CY				
Gamboge	Gam				
Pale Chrome	PC				
		Ivory Black	Bk		
		Snow White	SW		
		White Paper (Pirie, Aberdeen)			

there were different tints representing different degrees of saturation of the various hues. This too was a major innovation over his predecessors who, because of painting their colours, required a new disc for any change of colour.

FIRST EXPERIMENTAL RESULTS

A first example of his procedure was given by Maxwell in terms of making a neutral grey from vermilion, emerald green and ultramarine. The coloured papers were placed on the disc with the smaller papers, snow-white and black on top of them. The object then was to form a matching neutral grey following Maxwell's prescription for operating the top:

> the most perfect results are obtained by two persons acting in concert, when the operator arranges the colours and spins the top, leaving the eye of the observer free from the distracting effect of the bright colours of the papers when at rest. [5, pp127–8]

The operator, of course, was Maxwell himself. By suitable adjustments he was soon able to find that the outer and inner circles were 'perfectly indistinguishable when the top has a sufficient velocity of rotation.' [5, p128] The number of divisions occupied by each colour sector was then read off the scale and an equation formed. The result was

$$0.37\ \mathrm{V} + 0.27\ \mathrm{U} + 0.36\ \mathrm{EG} = 0.28\ \mathrm{SW} + 0.72\ \mathrm{Bk}. \qquad (8.1)$$

The total on each side is 1.00 or 100 divisions of the scale. Notice that the black sector is much larger than the white which implies that the grey formed was a rather dark one; therefore these colours produced a grey of low luminance.

However, other, brighter greys can be produced by using paler colours from the selection in Table 8.1. Maxwell did precisely this in a methodical manner. Maintaining two of his standard colours on the top he substituted for the third in succession 'one of the colours which stand under it.' [5, p128] Hence, he kept two variables constant (although not their areas exposed) and substituted for the third. In this case, for example, a brighter grey was

$$0.34\,\mathrm{PC} + 0.55\,\mathrm{U} + 0.12\,\mathrm{EG} = 0.37\,\mathrm{SW} + 0.63\,\mathrm{Bk}. \tag{8.2}$$

It is certainly possible, of course, to perform experiments where the object is to produce a definite colour rather than grey. However, one must be extremely careful then to ensure that the two colours are not simply varieties of each other but are, in fact, identical. [5, p129] An example of such a *colour* match as opposed to grey matches was

$$0.39\,\mathrm{PC} + 0.21\,\mathrm{U} + 0.40\,\mathrm{Bk} = 0.59\,\mathrm{V} + 0.41\,\mathrm{EG}. \tag{8.3}$$

That is, a darkened combination of pale chrome and ultramarine was equivalent in appearance to a red and green combination. Such experiments are difficult to perform however. There is an

> inability of the observer to express the difference which he detects in two tints which have, perhaps, the same hue and intensity, but differ in purity; and also from the complementary colours which are produced in the eye after gazing too long at the colours to be compared. [5, p129]

Thus both a linguistic barrier and fatigue-induced complementary colours arose in these trials.

Notice that in the first two equations the colour matching was in terms of grey. The left side of the equation was a grey composed of three colours, while the right side was a grey composed of white and black. But grey may be regarded as a white of low luminosity as was shown by Forbes in 1849. Therefore, those equations represent not so much colour matches as in equation (8.3) but luminance matches of low luminosity white. This is consistent with Maxwell's requirement that the three standard colours must give white when combined in correct proportions. From this point of view, then, Maxwell's colour top must be viewed as a photometer, albeit of very restricted use, and his colour equations must be considered to be derived from photometric data. As stated before, Maxwell's method effectively executed Grassmann's thought experiment. But Pierre Bouguer at the very inception of photometry, had pointed out that brightness matching was quite difficult when the lights were not the same colour, because two kinds of matches were required, colour and luminance. In terms of adjustments to the colour sectors, the equality of the hues could well be masked by slightly differing luminances or vice versa. No wonder then that Maxwell encountered linguistic problems in these trials. The observer, contending with two variables simultaneously, would find considerable difficulty in ascribing an apparent non-match either to hue or to brightness and hence would have difficulty in expressing directions for change to the operator. Note also in equation (8.3) that black appears on the left side where it reduces the luminance of the pale chrome and ultramarine combination. Considering these difficulties most of Maxwell's colour equations in his first paper are in terms of grey which are both easier to perform and are more accurate. Nevertheless actual colour matches, as such, are not impossible nor excluded in principle. It was simply practical to avoid them when possible, an example of Maxwell following his own advice on the 'judicious selection of what is to be observed.' We shall also see in Chapter 9 that equations achieved with his colour box are also almost all in terms of white.

Let us now return to the three equations shown. The question is: are these equalities intrinsic to the colours themselves or do they actually relate to the constitution of the eye? If they are intrinsic equalities they must remain equal under different lighting or observing conditions and therefore the question is answerable by direct test. In daylight Maxwell produced the following match, using carmine instead of vermilion:

$$0.44\,\mathrm{C} + 0.22\,\mathrm{U} + 0.34\,\mathrm{EG} = 0.17\,\mathrm{SW} + 0.83\,\mathrm{Bk}. \qquad (8.4)$$

However, the same colours illuminated by gas-light (a rather yellowish light) produced

$$0.47\,\mathrm{C} + 0.08\,\mathrm{U} + 0.45\,\mathrm{EG} = 0.25\,\mathrm{SW} + 0.75\,\mathrm{Bk}. \qquad (8.5)$$

Considering yellow as a combination of red and green we would expect more reflection from the carmine and emerald-green, and less reflection from the ultramarine (blue). This is in fact exhibited by equation (8.5) compared to equation (8.4) and therefore indicates the effect of the yellowish gas light on the match. Moreover Maxwell observed the colour match of equation (8.3) separately through blue, yellow and red glasses. In each case the colour became mismatched in terms of hue and one side of the equation (i.e. the photometer) appeared brighter than the other. Therefore, these matches are not intrinsic equalities but are related to the eye itself, the only other possibility.

MAXWELL'S COLOUR CONSTRUCTION

All the predecessors of Maxwell were deeply interested in some kind of graphical representation of colour mixtures. Part of this interest, beginning with Newton, involved early gropings toward a clear concept of colour variables. These attempts, we have seen, were largely in terms of presumed primary colours although the first explicit enunciation of these variables came from Grassmann in other terms. In the absence of explicit statements about colour combinations they represented an organising process by which colour data could be presented for later reference. Rather than refer to columns of data such as Richard Waller's or to pages of equations such as Maxwell might produce, it was far simpler to present the data graphically. Within this tradition, Maxwell was no exception. Indeed, he produced more and better data than anyone previously and so had most need of a clear representation of his results. Besides being simply a graph of his data, however, Maxwell's colour construction also served as an analogue computing mechanism by which he predicted new colour equations.

Each of the three standard colours, vermilion, ultramarine and emerald-green, is placed at a vertex of an equilateral triangle. The equilateral property is for convenience only as any triangle will serve. (Maxwell here emphasised once more that the three colours were chosen for convenience only [5, p131] although it seems clear as said before that this was only to avoid controversy over a new and unpopular idea.) In terms of representing colours as mass points and finding centres of gravity, all colours composed of these three must be points inside the triangle; or, any point inside the triangle represents the centre of gravity of some combination of vertex colours. Following Maxwell, let us use the example of the experiment represented by equation (8.1). The object is to locate the position of neutral grey composed of vermilion, ultramarine and emerald-green.

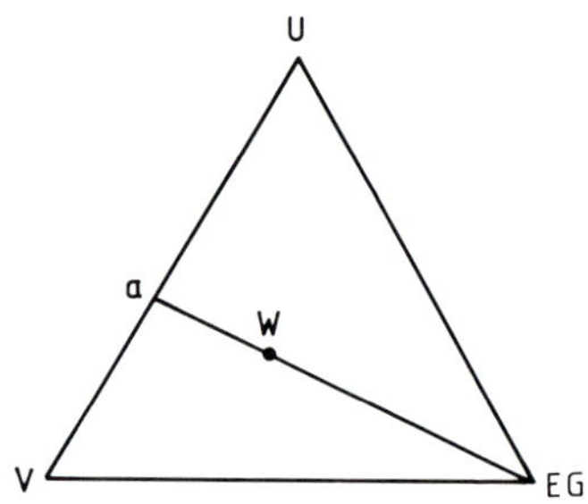

Figure 8.4 Locating white from Maxwell's colour equation (8.1).

In Figure 8.4, one imagines a mass of 0.37 units at V, one of 0.27 units at U and one of 0.36 units at EG. Taking the side UV arbitrarily one finds the centre of gravity between U and V. Calling this point 'a' and knowing the length of a side of the triangle, its position is determined from the proportion aV/aU = 0.27/0.37. Now simply repeat this procedure with the emerald-green. The centre of gravity of a line aEG will be called W and is found from the proportion

$$\frac{aW}{WEG} = \frac{0.28}{(0.37 + 0.27)}.$$

This point is shown in Figure 8.4, labelled W for white although the impression was actually grey. Ideally, however, the red, green and blue should give white, but, because of reflectance losses, the sensation is grey and this was the point of Forbes' analysis. To get reflected light of the same luminance as spectral colours, Forbes had shown that the proportions of his colours had to be increased inversely as their reflectances. A similar idea was now employed by Maxwell. He noted that the point W 'is not white, but 0.28 of white, diluted with 0.72 of black which has hardly any effect whatever except in decreasing the amount of the other colour.' [5, p132] What is desired, however, is a full circle of white, or 1.00 SW, not the grey represented by 0.28 SW. This represents ideally perfect reflection from the papers and bright white instead of grey. Therefore for the purpose of future calculations, Maxwell employed a kind of normalisation procedure in which one asks: What coefficient, X, is necessary to multiply 0.28 in order to give 1.00? The answer is $1/0.28 = 3.57$. This number represents a correction factor which normalises the 'white' measured under reflection to the equivalent of a white made by spectral colours. However, it must be realised that the normalisation is with respect to the three standard colours chosen U, V and EG which, by definition, have correction factors of 1 exactly. Maxwell spoke of the number as a coefficient to be used so that 'wherever white enters into an equation, the number of divisions must be multiplied by the coefficient 3.57 before any true results can be obtained.' [5, p132] We shall see shortly that this concept allowed Maxwell to make a great number of colour equations and perform analyses with the data obtained.

Let us now consider Maxwell's second experimental match exhibited in equation (8.2). The problem here is that pale chrome cannot be made from the standard colours. If all colours compounded of the three standards are points inside the triangle, where then are the colours not compounded from them? We saw that all compound colours in Newton's colour circle lay within the circle and he did not consider colours outside the circle. Presumably such colours had no physical significance. The same tacit assumption was involved in Mayer's triangle. A three-dimensional array was really required as was implied by Mayer's and Lambert's pyramids. It was here that Maxwell surpassed his predecessors again for we shall now see that PC can be plotted with respect to the triangle of colours using the coefficient concept above. This is the object of the construction that follows.

Now whenever white enters an equation we multiply its experimental value by the coefficient to obtain an 'ideal' white. Hence, from equation (8.2) we have $(0.37)(3.57) = 1.32$. It is now 1.32 which is to measure the total brightness on both sides of the equation. (There cannot, in principle, be any corrected values of the standard colours, U and EG.) Therefore we subtract the brightness of 0.55 U + 0.12 EG (=0.67) from 1.32 and obtain a value of 0.65 for pale chrome. This is a corrected or normalised value of PC and is to be regarded as an ideal value in terms of the ideal white. We can now rewrite the equation (using lower case letters instead of capitals for corrected values) as

$$0.65\,\mathrm{pc} + 0.55\,\mathrm{U} + 0.12\,\mathrm{EG} = 1.32\ \mathrm{w}.$$

This equation means that the papers are ideal reflectors and that the top has 132 divisions instead of 100. It is now easy to locate the pale chrome. Since we know the positions of U and EG we begin by finding their centre of gravity (see Figure 8.5). Calling this point β, it is determined from the proportion

$$\frac{\beta \mathrm{U}}{\beta \mathrm{EG}} = \frac{0.12}{0.55}.$$

Now since white is the centre of gravity of the combinations its position is

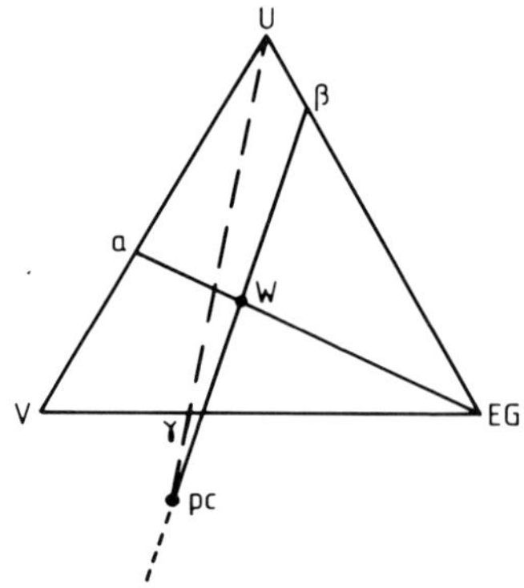

Figure 8.5 Locating the position of pale chrome in terms of the standard colours from Maxwell's colour equation (8.2).

already known, the position of pc must lie on a line from β through w and on the other side of w from β. This point is determined by

$$\frac{\text{pcw}}{\beta\text{w}} = \frac{0.55 + 0.12}{0.65}.$$

By the use of the coefficient for white we have now been able to locate a colour *outside* the triangle. For the first time colours had been located outside the triangle thereby giving colours not specifically composed of the primaries a physical interpretation. As shown in Chapter 6, Newton's circle or the triangles of Mayer and Lambert could be regarded as exhibiting two of Grassmann's variables, hue and saturation, with brightness held constant. Inclusion of brightness required the third dimension of space. With his normalisation procedure Maxwell has effectively succeeded in plotting three-dimensional data in two dimensions. This concept was crucially important to the emergence of the modern chromaticity diagram to be discussed in Chapter 9. Note also that the procedure has given a coefficient for the pale chrome paper. Observed values of pale chrome must be corrected in the ratio 65/33 or 1.97. [5, p133]

There is still the question, however, as to whether this 'method of coefficients' is really valid. If so, then knowing the positions of the coloured papers with respect to the triangle, we should be able to reason in reverse from the diagram to the top. We should now be able to measure off on the diagram correct proportions for colour matches, and these predicted results should agree with measured values on the top. Maxwell followed this procedure to test the experimental results of equation (8.3). We have to see if these results are achieved in the diagram. We join U with pc, and V with EG (already done in the triangle). These lines intersect at the point γ. But from the diagram γ is the centre of gravity between V and EG and between U and pc. By simply measuring the proportions of the line VEG, Maxwell found

$$0.58\,\text{V} + 0.42\,\text{EG} = 1.00\gamma.$$

on the line Upc Maxwell found the proportion, *by measuring*, of

$$0.78\,\mathrm{pc} + 0.22\,\mathrm{U} = 1.00\gamma.$$

But the value of pc found here must be its corrected value. Since the normalised values were found by multiplying the observed values by the coefficient, here we divide the corrected value by the coefficient for the pale chrome to get the observed value. Therefore we have 0.78pc/1.97 = 0.39PC. We can now equate the two equations for the point γ, substituting the observed value of PC and the result is almost exactly as before in equation (8.3):

$$0.39\,\mathrm{PC} + 0.22\,\mathrm{U} + 0.39\,\mathrm{Bk} = 0.58\,\mathrm{V} + 0.42\,\mathrm{EG}.$$

The black as noted previously acted as a darkening agent reducing the brightness of the left-hand side. The implication is that the combination of 0.39 PC + 0.22 U has the same apparent hue as 0.58 V + 0.42 EG but is brighter. For the same colour sensation, which depends on *both* hue and luminance, the black is necessary to darken the PC and U combination giving luminance matching as well, a photometric requirement of this procedure. As Maxwell said the black was used 'simply to fill up the circle' [5, p134] which is legitimate since black is no colour at all, indeed being no light at all ideally. Note that the triangle was used to solve graphically two simultaneous equations. An equation was made of each side of equation (8.3) with the colour at γ representing the common solution of the equations.

The important point is that the values measured from the diagram are very close to those measured from the top, the difference being only one per cent. This small difference lay well within the range of error obtainable by the top. Of experiments with 10 observers in November 1854 Maxwell stated that:

> the best were accurate to within 1–1/2 divisions, and agreed within 1 division of the mean of all; and the worst contradicted themselves to the extent of 6 degrees, but still never more than 4 or 5 from the mean of all observations. [5, p130]

Hence predictions from Maxwell's colour construction agree well within the observational limits and the 'method of coefficients' is therefore valid. Maxwell's colour diagram for all the papers is shown in Figure 8.6. The number beside each colour is the coefficient for that paper.

From the experiments with 10 observers and the results illustrated in Figure 8.6, Maxwell concluded the following:

> 1st That the human eye is capable of estimating the likeness of colours with a precision which in some cases is very great.
>
> 2nd That the judgement thus formed is determined, not by the real identity of the colours, but by a cause residing in the eye of the observer.
>
> 3rd That the eyes of different observers vary in accuracy, but agree with each other so nearly as to leave no doubt that the law of colour vision is identical for all ordinary eyes. [5, p130]

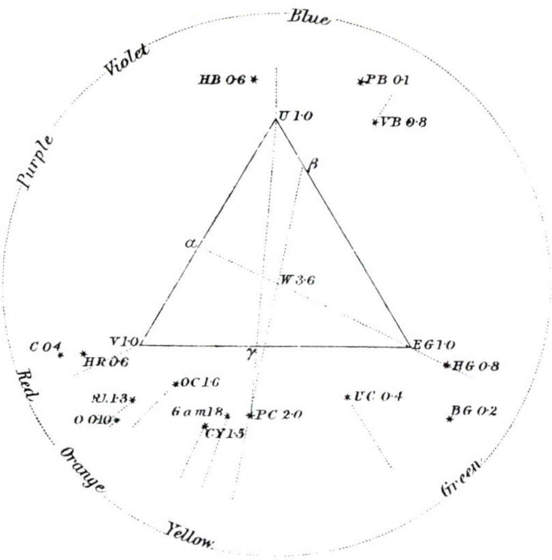

Figure 8.6 Maxwell's location of all his coloured papers given in Table 8.1. He has not drawn a curve through the plotted points but only indicated that the locus is roughly circular in the sense of Newton.

The third conclusion answers the question mentioned before as to whether everyone's colour sensations are the same. We would certainly allege an affirmative answer from Grassmann's linguistic arguments. The very existence of a vocabulary of colour words and terms upon which there is general agreement indicates a positive answer. The amazement expressed by normal people at the mistakes of the colour blind also argues an affirmative answer. But now the question had actually been put to a strict empirical test by Maxwell leaving no doubt that perceived colours yield essentially equivalent sensations for everybody. (This was put to an even more stringent test by Maxwell in 1860 with a slightly amended conclusion.) This must certainly be an early example, if not the first, of a psychological question receiving a direct empirical test. Interestingly enough Helmholtz, shortly afterward, provided a similar conclusion for sound impressions in his analysis of the ear and hearing.

MAXWELL'S COLOUR CONSTRUCTION AND GRASSMANN'S COLOUR VARIABLES

So far Maxwell had employed a theory of three colour sensations involving red, green and blue. But Grassmann had proposed three colour sensations involving luminance, hue (wavelength) and saturation. Although they agreed on the number of variables involved, it appeared that they disagreed on exactly what they were. Grassmann's variables inferred from observational and linguistic arguments were strongly reasoned. They had the additional virtue of avoiding any controversy over primary colours. The choice of three primary colours, alternatively, was less strongly based having no known anatomical basis and having undergone no really rigorous test. Not even Helmholtz's work of 1852 was *de facto* evidence for this idea since he himself had wondered whether perhaps five colours would not be needed to imitate all the spectral colours. His work of 1853, however, would certainly argue more strongly for three primary colours. The question being open nevertheless, Maxwell undertook to show that no disagreement in fact existed between the two sets of variables and that 'the two methods of considering colour may be deduced one from the other, and are capable of exact numerical calculation.' [5, p131] In order to avoid confusion between the different terminologies used by Grassmann and Maxwell let us first note Maxwell's nomenclature and his reasons for it.

The first variable Maxwell mentioned was 'shade'. That is, considering two objects of the same apparent hue, one could be brighter or darker than the other. The second variable mentioned was 'hue', or simply the specific name of the colour. Finally, considering two samples of the same colour, Maxwell noted that they could 'vary from purity on the one hand, to neutrality on the other.' [5, p131] This variation he called 'tint'. Therefore the other set of colour variables for Maxwell were shade, hue and tint, which are shown below next to Grassmann's terms together with their modern equivalents.

Maxwell	*Grassman*	*Modern*
shade	brightness	luminance
hue	hue	dominant wavelength
tint	intensity of intermixed white	purity, or saturation

The identity of Maxwell's construction with Grassmann's variables is now easy to see. In Figure 8.6, a series of lines drawn from W to each of the colours plotted passes through all the colours of the spectrum taken in order. Also note that W lies near the centre of the triangle while the most saturated colours lie toward the circle, or far from the centre of the triangle. (W is in fact the centre of the circle.) We conclude therefore that hue is determined by angular position around the circle and tint (i.e. saturation) is determined by the distance of the colour point from W. The brightness of the colour is specified by its coefficient which in turn must be determined to construct the diagram. Therefore, the three variables first identified by Grassmann are now formally identified on this diagram although the diagram itself is based on the assumption of three primary (standard) colours. This being so, Maxwell concluded that 'the relation between the two methods of reducing the elements of colour to three becomes a matter of geometry.' [5, p135] Maxwell's construction compared with Newton's colour circle (Figure 4.1), shows that they are essentially the same; thus Maxwell's colour equations and his method of coefficients enabled him to produce the first *empirical* derivation of Newton's colour organisation. In doing so he had also carried out Grassmann's aim, and the genius of Newton was confirmed through the genius of Maxwell.

Maxwell's emphasis on a three-fold process of colour vision is most interesting in terms of Thomas Young's theory of vision. That theory we know had been essentially groundless at its inception. Proposed within the context of his undulatory theory, it had been added, as that eminent authority 'Pooh-Bah' might say, 'to give artistic verisimilitude to an otherwise bald and unconvincing narrative.' [16] Young had simply seized a popular notion about three primary colours and elevated it to the status of formal scientific doctrine. In principle, of course, this would have been quite reasonable if he had made strong arguments in its favour from empirical observations. In general, a criterion for acceptance of an hypothesis, at least for discussion purposes, is that it seems reasonable. Thus Grassmann argued convincingly from different sources of evidence that colour vision involves three variables. This process, of course, was not followed by Young who proposed his speculation within the framework of yet another initially unconvincing theory. He left it for others to marshal the necessary evidence. Thus for want of evidence and for want of fundamental arguments, Young's hypothesis of three-colour vision withered. But if it withered, it did not die. It was just provoking enough to elicit comments from time to time, as we have seen, from people such as Brewster, J Herschel, Forbes, Helmholtz, Wartmann, and others of lesser fame. For Young's hypothesis, the problem of acceptance involved gathering the necessary empirical observations to make it look respectable. Helmholtz initiated that process with his V-slit experiments, but they themselves were not sufficient evidence. The process of adding evidence was continued by Grassmann but that famous paper while convincing had the disadvantage, perhaps, of being too intuitive and too theoretical. Again it was insufficient in itself or even together with Helmholtz's subsequent work. What was required was some hard quantitative data and that was finally supplied by Maxwell. The experiments described above and Maxwell's construction were his initial contribution to the process of placing Young's three-colour speculation on an empirical basis. They also, of course, indicate Maxwell's deep commitment to the idea. We shall shortly examine more of Maxwell's experiments involving colour blindness which give additional credence to Young's ideas. However, we shall pause first to examine Maxwell's very graphic illustration of Young's theory.

THE FIRST COLOUR PHOTOGRAPH

Within the context of illustrating the three-fold nature of colour vision and as a simulation of Young's theory, Maxwell proposed in this paper of 1855 nothing less than a method of colour photography. In the infancy of photography, when producing black and white photographs posed problems enough, Maxwell prescribed the modern steps for producing colour separation negatives.

His idea was elegantly simple. Using a black and white negative emulsion 'equally sensitive to rays of every colour' [5, p136] the process is as follows. (Note that whether or not such an emulsion was available at the time was not a consideration; it was simply a proposal of principle within the paper.) Employing one plate, photograph a scene through a red filter. Then a *positive* plate from this negative will be opaque where the scene reflected little red light and transparent where the scene reflected much red light. Placing this positive in a magic lantern projector with the same red filter before the lens will then give a projected image of those parts of the scene that were red. Now repeat the operation with a green filter and again with a blue filter. Then the three positive plates are placed in three separate magic lanterns each with its corresponding filter before the lens. Upon superimposing the images a coloured image of the scene must be produced including intermediate colours compounded of these primaries. Although the anatomical nature of three-fold vision remained unknown (and still does) Maxwell stated that the illustration would demonstrate that 'the functions which Young attributes to the three systems of nerves may be imitated by optical apparatus.' [5, p137] It was an ingenious proposal.

Most amazingly in 1861 Maxwell actually achieved this result at the Royal Institution while giving a Friday Evening Discourse on colour. Plate 8 is a reproduction of the photograph from Maxwell's own plates. The *Record of the Royal Institution* recorded the result as follows:

> three photographs of a coloured ribbon taken through the three coloured solutions respectively, were introduced into the camera, giving images representing the red, the green, and the blue parts separately, as they would be seen by each of Young's three sets of nerves separately. When these were superposed, a coloured image was seen, which, if the red and green images had been as fully photographed as the blue, would have been a truly-coloured image of the ribbon. By finding photographic materials more sensitive to the less refrangible rays, the representation of the colours might be greatly improved. [17]

Now although the concept was technically flawless it was realised some years later that emulsions then available were *completely* insensitive to red, orange, yellow and green. Their narrow sensitivity ranged only from the extreme blue at about 430 nm to the near ultraviolet at about 330 nm. Therefore the demonstration should not have worked! But the separation plates had been prepared in good faith by Thomas Sutton, a well known lecturer on photography; they were exhibited by Maxwell in similar good faith and were seemingly sufficiently convincing. Thus arose the question as to what unknown factors were operating when Sutton made the plates that actually did separate the colours. It will be interesting to take a short tangent here to see the solution to this puzzle. This was produced by Ralph M Evans of the Colour Technology Division of Eastman Kodak who repeated Maxwell's (and Sutton's) experiment according to Sutton's records using copy positives of Maxwell's original plates then still at the Cavendish Laboratories. [18]

The explanation for the red plate turned out to be easy. It is now known that many red dyes reflect ultraviolet light as well as red. Evans' reflectance curve for a red cloth shows almost 20% reflectance of ultraviolet at about 370 nm. Moreover, the red filter Sutton used, a solution of ferric thiocyanate, exhibited similar transmission characteristics while the glass of Sutton's lens, it was established, had an ultraviolet cut-off at about 300 nm. Therefore, Sutton's 'red' separation plate was really an ultraviolet separation from ultraviolet reflection by the red dye in the ribbon. It was as Evans says 'a happy accident that neither Maxwell nor Sutton could have foreseen.' [18, p125] This explanation also solved another puzzle. Of Sutton's three plates, the red one was markedly less sharp than the others although evidence from the plates themselves shows that he refocused for each colour. It is clear now that he focused his camera in red light but photographed in ultraviolet which produces a shorter image distance than the red.

A slightly different explanation arises for the green separation plate. Sutton had used a cupric chloride solution for this filter. But with a good green solution he was originally unable to obtain a negative after a twelve minute exposure, a very long exposure even then. Accordingly he diluted the solution until he was able to produce 'a tolerable negative' [18, p118] after twelve minutes. It turns out, however, that the actual colour of cupric chloride solutions depends on their concentration. Sutton's final filter giving the tolerable negative was he said 'very much paler' [18, p118] than the original solution. Repeating this part of Sutton's work, Evans found that his reproduced filter had turned from deep green to bluish-green when he finally achieved Sutton's result. [18, p122] Also his transmittance curve for this filter indicates a pass band that just overlaps the upper cut-off limit of the film at about 430 nm. Therefore the 'green' separation plate apparently resulted from blue light at about 450–430 nm reflected from the green portion of the ribbon. Now we can understand why the red and green images were not as 'fully photographed as the blue'. Some very fortuitous circumstances allowed Maxwell to produce the first colour photograph some 45 years before it became technically possible.

It was nevertheless in 1861 a very dramatic argument for the three-fold nature of colour vision. The original proposal in the 1855 paper, following as it did the experiments and construction already described, was intended to produce the same effect. As already noted persuasion as well as logic is needed in order to change a paradigm. Maxwell's proposal was surely intended in just this way regarding the old idea of yellow rather than green as a primary colour. The proposal and its later realisation were to demonstrate the validity of the three primary colours, red, green and blue, together with Young's original hypothesis that other colour sensations are compounded from these three. There was, however, yet more evidence available for the nature of this process, this time from colour blind subjects producing colour matches on Maxwell's top. The experiments are intrinsically very interesting, adding important corroboration to Young's trichromatic theory, as well as to Maxwell's triangle construction.

COLOUR EQUATIONS AND COLOUR BLINDNESS

The last chapter traced the efforts at identifying colour blindness through the first half of the nineteenth century. These efforts culminated in the work of Wilson in which Young's theory of vision was tentatively espoused. But all this work was qualitative, depending very much on colour description by the affected individuals. No quantitative tests had been devised to eliminate the communication problem between subject and investigator. The best approach in this direction had been the work of Seebeck from which the concept of blindness to red had arisen, although the idea can be traced as we saw to Dalton and Young. Seebeck had shown that of his two classes one could hardly see the red end of the spectrum while the other could see it fairly well. However, this opinion was somewhat clouded. Dalton, who presumably saw little or no red, was tested in this way by Brewster who declared, contrarily, that Dalton saw a spectrum of proper length and hence did see red. Brewster had invoked this as additional proof for his triple spectrum. Therefore although Young's idea of a missing red sensation (to explain Dalton's vision) was a convenient one there was certainly quite some confusion over just what colour blind people saw. A rigorous quantitative test of Young's concept was needed to eliminate the necessity for colour description by the subject. Maxwell supplied such a test in these first colour vision researches.

Among his subjects, Maxwell found four who were decidedly colour blind and he reported results on two of them. These persons were asked to form colour matches with the colour top in much the same manner as normal persons. However, there was a significant difference. In experiments with normal people the operator (Maxwell) could conceivably influence the results since he received essentially the same impressions as his subject. Hence in questioning his subject for adjustments, as already described, his phrasing could be influenced by what he himself saw. But in testing the colour blind this was impossible since the operator and the subject had distinctly different impressions. The operator could now ask only two questions: 'Which circle appears most blue or yellow? Which appears lightest and which darkest?' [5, p138] The latter question is the photometric one for equal luminance between the two colour circles. The first question is reminiscent of Dalton's self-description in which he said he saw only two colours in the spectrum, yellow and blue. Thus a truly red blind person must be asked only about these colours and then only in the terms in which he sees them. Under these conditions Maxwell obtained colour matches and hence colour equations from his two colour blind subjects. One such match using red, green, blue and black papers was as follows, where the values given are the mean of several observations:

$$0.19\,\mathrm{G} + 0.05\,\mathrm{B} + 0.76\,\mathrm{Bk} = 1.00\ \mathrm{R}. \qquad (8.6)$$

The first interesting point here is that only four colours including black are needed for a match rather than five as in the preceding equations. Ignoring the black momentarily, it appears that the combination of green plus blue is equivalent to a red. The black then acts as a darkening agent reducing the brightness of the green–blue combination. Therefore to this subject the equation shows that a dark green–blue is equivalent to a full red. [5, p138] It is perhaps better said that the red blind cannot discriminate between these colours which is confirmed by an anecdote concerning William Whewell and Dalton. At a Cambridge University ceremony Whewell asked Dalton what other colour his Doctor's gown, a bright scarlet, resembled. Dalton pointed to some evergreens outside the window remarking that to him their colour and that of his gown were alike. However, the lining of the gown was pink which Dalton said he could not distinguish from sky blue. [19] The latter observation, of course, is exactly the same as his 'encounter' with the geranium.

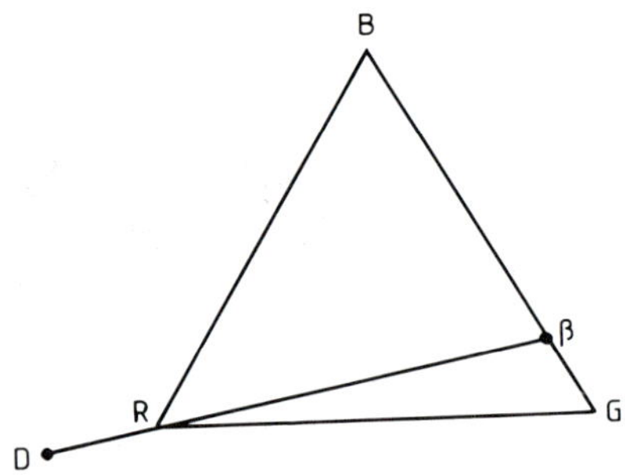

Figure 8.7 Locating the missing red sensation, point D, according to Maxwell's methods and his colour equation (8.6). It is analogous to black for this observer.

Now Maxwell applied his colour triangle method to this equation using a new diagram since he had a different set of papers. In Figure 8.7 the standard colours (Hay's papers) red, green and blue are located at the vertices. As before we locate the centre of gravity of green plus blue imagining weights placed at these vertices of amounts measured on the top. Calling this point β, the line BG is divided at β in the proportion of 19:5. Then β is joined with R as shown. Now a new insight into the vision of colour blind people becomes possible. The equation shows that the hue at β appears essentially the same to the colour blind as that at R. It is simply brighter than R in the proportion of 100:24, or almost 4:1. (This arises from the amount of green plus blue required to match the red. The red represents a full circle, or 100 divisions; 19 divisions of green and 5 divisions of blue were much brighter to the subject than the red and needed to be darkened to produce the match. They were in fact almost four times brighter.) Not only is the point β essentially the same as red, but the line Rβ must represent colours to the colour blind which are all the same hue but differing only in brightness. As Maxwell said, 'all intermediate tints on the line Rβ will appear to them of the same hue, but of intermediate intensities.' [5, p139]

This leads to a further insight. Suppose some colour other than black could be added to equation (8.6) on the left-hand side to match the red. The equation could be written as

$$0.19\,\mathrm{G} + 0.05\,\mathrm{B} + 0.76\,\mathrm{X} = 1.00\,\mathrm{R}$$

Where then would one locate the unknown colour X? Following the centre of gravity method the question can be answered. The centre of gravity of the three colours on the left must be at R. The centre of gravity of B and G has already been located at β. Therefore, X must lie on the other side of R, outside the triangle and is shown in Figure 8.7 as point D. Point D is located by the proportion RD/Rβ = 24/76. D must represent a sensation to the colour blind which is invisible! Assuming a paper existed of the colour represented at D it would represent a specific sensation unknown to a colour blind person. An unknown sensation is no sensation; for *him* it is equivalent to black. In terms of Young's theory, Maxwell had located geometrically the missing sensation for the red colour blind. The actual optical wavelength was to be determined at a much later date.

Figure 8.8 is Maxwell's complete drawing for all the data of this experiment. The positions of white and yellow were determined from normal observers. On the diagram Maxwell has drawn the lines DW, DB, DG, etc.

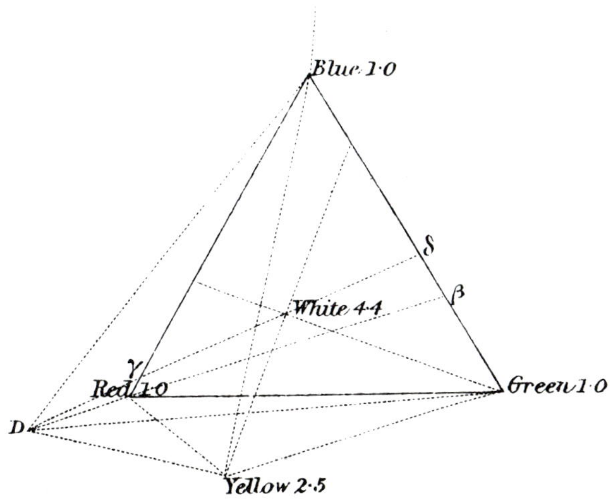

Figure 8.8 Locus of colour blind colour matches as observed by Maxwell's colour blind subjects. Point D represents the missing red sensation.

These lines are analogous to the line Dβ. Along that line, starting from R, the subject saw all colours as the same hue but of varying brightnesses. Similarly, for the colour blind 'the colours belonging to any line ought to differ in intensity as seen by them, so that one of them may be reduced to the other by the addition of black only.' [5, p139] Moreover, the line DW, drawn through white as composed of red, green and blue, and produced to δ is a boundary line for the colour blind. Since he does not see the red represented at D, all colours above Dδ must be varieties of blue; those below the line Dδ must be varieties of yellow. Dδ is a boundary between the two colours seen by Dalton, or Seebeck's Class II; to these people all colours appear as if composed of yellow and blue.

The confusion about a missing sensation or the degree of red visible to such people was now cleared. The paper at R is a red and is obviously seen by this person. Hence he does see red to some extent. But the normal eye experiences other 'red' sensations out to the position D which this person cannot see. Therefore the red sensation is markedly truncated for the red blind person. In effect the standard colours for the colour blind subject are green, blue and black. This was already noted in Figure 7.1 which shows in modern terms that the threshold of vision for someone like Dalton is shifted sharply towards the orange. Thus this diagram represents a first attempt at spectral sensitivity measurements. Chapter 9 discusses Maxwell's further development of this idea using spectral colours.

So far Maxwell had provided strong evidence in favour of trichromatic vision. Perhaps the strongest argument elicited by his investigation was the demonstration that the colour blind lacked one colour sense just as expected on that theory. However, the foregoing data and interpretations were subject to additional confirmation. With normal subjects, Maxwell had predicted colour matches from measurements on the colour construction. The construction (Figure 8.8) is basically the same as the first, having been formed originally with normal eyes. However, the points D, β and δ were determined from the colour blind data so the construction now exhibits both classes of data. Hence, it should be possible from this to *predict* colour matches for the colour blind, a new and unusual approach to understanding their vision. This Maxwell did with results which form the keystone to a new understanding of colour vision.

In Figure 8.8, notice that the line DW crosses RB at γ. By measurement on the triangle we find that $R\gamma/B\gamma = 0.93/0.07$, or that

$$0.07\ B + 0.93\ R = 1.00\ \gamma.$$

Similarly, measuring on the line DW we find that the ratio of $W\gamma/D\gamma = 0.895/0.105$, or that

$$0.895\ D + 0.105\ W = 1.00\ \gamma.$$

Thus, as before, the construction creates two simultaneous equations. We equate these two but note first that the point D was shown to be invisible to the colour blind. In equating these two (see below) the letter D, following Maxwell's example of the 'colourblind language', is now called black. The measurements predict the following colour blind match which is shown with measured values underneath it.

predicted	$0.93\,R + 0.07\,B = 0.105\,W + 0.895\,Bk$,
observation of two	
subjects together	$0.94\,R + 0.06\,B = 0.10\,W + 0.90\,Bk$,
one subject alone	$0.93\,R + 0.07\,B = 0.10\,W + 0.90\,Bk$.

The percentage difference is just over 1% which was previously shown to be within the limits achievable from the colour top. One other example will show the power of this method. Extending the line DW to δ gives a similar equation between points D, δ, B and G. The predicted results from the triangle and those observed are shown below. [5, p140]

predicted	$0.43\,B + 0.57\,G = 0.335\,W + 0.665\,Bk$,
observation of two	
subjects together	$0.41\,B + 0.59\,G = 0.34\,W + 0.66\,Bk$,
one subject alone	$0.42\,B + 0.58\,G = 0.32\,W + 0.68\,Bk$.

Finally, other intersections in the figure give relations between R, Y, G, Bk; Y, B, R, Bk; and Y, B, W, Bk all predicted and confirmed by observation. Note that in each group here and above there are four colours including Bk and W.

These colour blind equations compared to Maxwell's normal observer equations (8.1)–(8.4) are most interesting for they provide strong evidence of a trichromatic system of normal vision. The abnormal eye needs only four colours including white and black to make colour matches rather than five as required by the normal eye. Hence, as predicted long before by Young, one of the colour sensations is absent, specifically red, for these cases of Maxwell's or Seebeck's Class II. Or, not counting white and black as colours, the normal eye requires three colours for colour matching while the red blind eye needs only two. Considering this, Maxwell said of Young's hypothesis that it 'affords a most trustworthy foundation on which to build a theory of colourblindness, as it expresses completely the observed facts of the case.' [5, p140] In fact, in Maxwell's hands it also furnished a new datum, point D, whose exact nature remained to be determined. This new quantitative view of Young's hypothesis also implied that people with absent green or blue sensation should exist. Presumably a constructional point analogous to D should be observed for them also. Such people do, indeed, exist and part of the later history of colour vision involves a search for such persons. Suffice it to say now that a point analogous to D has been observed for tritanopes or blue blind persons but measurements expected to lead to a similar point for deuteronopes, or green blind individuals, have given ambiguous results. [20] The problem is still an active one.

Maxwell also attempted to aid colour blind people in distinguishing colours. Wilson had shown that colour blindness was incurable and that coloured glasses hardly helped such people. Maxwell, however, pointed out that the subject of equation (8.6), who saw red and green as identical, could be aided by coloured glasses. A red glass gives a fairly clear transmission of red while suppressing green, giving it an obscure, dark appearance. A green glass similarly suppresses red while passing green. Therefore Maxwell thought that distinctions between red and green could be made on the basis of luminance by viewing colours through the appropriate glasses alternatively. The subject of equation (8.6) was indeed helped, so that in cases of doubtful judgement he was enabled to judge with certainty. This perhaps was not entirely new, but Maxwell's predecessors had used only one coloured glass in such attempts rather than comparing brightness with two different colours. Maxwell wrote to Thomson saying that 'I have enabled such a man to distinguish reds and greens by showing him the opposite effects of red and green glasses on colours similar to him.' [8] He even went so far as to construct a pair of spectacles with one green glass and one red glass, proposing that the two different filters used simultaneously would aid him in colour discrimination [5, p141] No further mention of these attempts appears in Maxwell's later works or correspondence so presumably the results were not as he expected.

RECAPITULATION

This work of Maxwell represents a culmination of various lines of research on colour perception (excepting photometry) all enriched and enlarged by Maxwell's genius. More contributions were to come which are discussed in Chapter 9. There they will be seen as refinements in technique together with enlarged insight into colour vision generally. In the meantime it is useful to have a short recapitulation here in order to see how the diverse investigations discussed in the earlier part of the story have now merged.

Figure 8.9 illustrates briefly the nature of this culmination. It does not necessarily imply that the work of one man listed above another directly stimulated the later investigator, it simply shows that theories of colour and colour perception can be viewed as developing along semi-independent lines. The double ended arrows indicate that none of these lines of inquiry are fully independent of each other and that, if anything, there has been a kind of cross fertilisation between them all. To investigate the history of any one of these avenues of colour knowledge is informative but incomplete without a view of the others.

Columns (1) and (2) indicate the facts described in Chapters 7, and 1, 2 and 3 respectively. The modern problem of colour perception was introduced by Young in the context of his wave theory of light in the Bakerian Lecture of 1801. But we have seen that he was probably stimulated in this idea by prior work of Newton and the general notion already prevailing of three primary colours. In the latter idea Young and Wollaston substituted green as a primary colour rather than yellow. Approximately contemporary with this was the first formal study of colour blindness by Dalton in 1794 and Young's comment on Dalton's sight made in 1807. Hence, we have an early crossing of ideas but still in a very speculative state. Brewster's triple spectrum must be viewed as a theory of vision since he assigned an objective reality to his

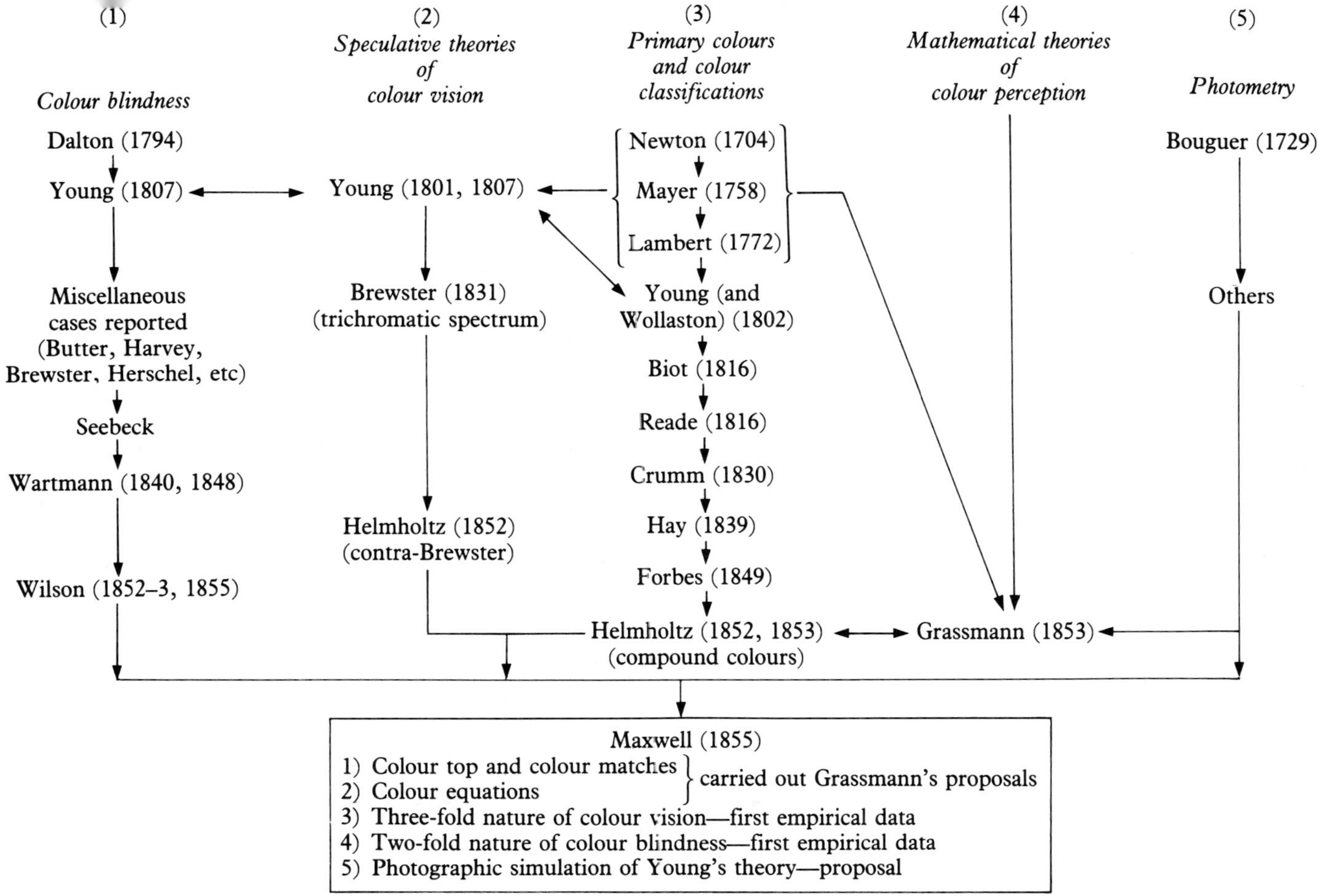

Figure 8.9 Development of research into colour vision culminating with Maxwell's theories.

three spectral primaries. It was even necessary for Helmholtz to refute it as a preliminary to his more important work on compound colours. Colour blindness investigations in column (1) could and did proceed independently of the other lines in terms of description of case histories. Still every scientist looks for an understanding of his empirical data beyond mere description; and so within this tradition there was also constant discussion of theories of vision, especially Young's. Column (3) does actually represent a somewhat linear line of development except that Lambert arrived at his colour pyramid independently of the work of Mayer. The arrow linking these attempts with that of Grassmann indicates that the works of Newton, Mayer and Lambert may also be viewed as early attempts at a mathematical understanding of colour; these attempts, however, were all in terms of classification of colours which subject Grassmann did not consider.

Helmholtz's papers of 1852 and 1853 represent a unification of these two lines in terms of Young's theory. It was, however, a qualitative union only, subject to the corrections and development given by Grassmann. Moreover, the subject of colour classification can be traced beyond 1852 well into the twentieth century. Still in terms of understanding colour vision at that time, Helmholtz's work represented a fusion of two lines of approach. Photometry, as a science, developed quite independently of all these subjects but Grassmann borrowed its ideas to indicate the possibility of performing colour matches with light. This borrowing is indicated by the single ended arrow. With the work of Helmholtz and Grassmann the stage was set for a quantitative exploration of colour matching by a photometric method, an analysis of the data according to Grassmann's vector concepts (or Newton's colour circle which is essentially the same idea), final acceptance and verification of Young's hypothesis of three-colour vision, and a quantitative explanation of colour blindness by Maxwell. All these aspects of Maxwell's work are seen below columns (1)–(5) and are intended to indicate that in Maxwell a great deal of earlier work crystallised and bore significant fruit.

Much has been written on Maxwell as a unifier of electrical knowledge and a founder of kinetic–molecular theory. This diagram and indeed this entire chapter show that Maxwell was also a unifier of knowledge on colour theory and colour perception. Moreover, he was an important innovator of quantitative experimental methods, the very thing this science lacked. (This will be seen especially in the next chapter.) It was Maxwell who implemented Grassmann's theoretical ideas with operational equipment. Helmholtz not withstanding, it was Maxwell who set the infant science on a new course by showing how to proceed. The mark of his success has been that a great deal of modern work stemmed from this foundation; his method of colour matching and his colour equations are still very much in use.

NOTES

1 Campbell, Lewis and Garnet, William 1882 *Life of James Clerk Maxwell* (London: Macmillan)
2 *Rep. Br. Assoc.* 1856. Reprinted in 1890 in *The Scientific Papers of James Clerk Maxwell* vol I (London: Cambridge University Press) p242
3 Merz, John Theodore 1923 *A History of European Thought in the Nineteenth Century* 4th edn vol 1 (Edinburgh and London) p271
4 Tait, P G 1879–80 *Proc. R. Soc. Edinb.* A long excerpt from this eulogy of Maxwell appears in [1, pp133–4] upon which this statement is based
5 Maxwell, James Clerk Experiments on colour as perceived by the eye, with remarks on colour blindness. Republished in 1890 in *The Scientific Papers of James Clerk Maxwell* vol I (London: Cambridge University Press)
6 For confirmation see also a letter from J D Forbes to Maxwell dated 4 May 1855 [1, pp213–14]
7 Maxwell discusses the equivalence of several other methods in [5, pp143–7]
8 Larmor, Joseph (ed) 1937 *Origins of Clerk Maxwell's Electric Ideas, as described in familiar letters to William Thomson* (London: Cambridge University Press) p7. Letter of 14 March 1854
9 Letter of 24 November 1854 to Miss Cay [1, p209]
10 Maxwell, James Clerk 1855 in George Wilson 1855 *Researches in Colour Blindness* (Edinburgh). The letter is dated 4 January 1855 and was solicited by Wilson who had earlier asked Maxwell for the results of his colour experiments. (Also printed in 1890 in *The Scientific Papers of James Clerk Maxwell* vol I (London: Cambridge University Press))
11 Letter of J D Forbes to Maxwell dated 16 May 1855 [1, p214]
12 Maxwell, James Clerk Address to the Mathematical and Physical Sections of the British Association 15 September 1870 Reprinted in 1890 in *The Scientific Papers of James Clerk Maxwell* vol II (London: Cambridge University Press) p217
13 Kuhn, Thomas 1962 *The Structure of Scientific Revolutions* (Chicago: University of Chicago Press)
14 Letter dated 24 September 1855 [1, p216]
15 The necessity for this process is argued by Kuhn [13, Chapter IX]
16 Gilbert, W S *The Mikado*. Line spoken by the character Pooh-Bah in the second act
17 *Record of the Royal Institution*. Reprinted in 1890 in *The Scientific Papers of James Clerk Maxwell* vol I (London: Cambridge University Press) p449
18 Evans, Ralph M 1961 Maxwell's Colour Photograph *Sci. Am.* November 118–27. See also *J. Photog. Sci.* **9**
19 The anecdote was related by Whewell himself from the floor at the 1841 meeting of the British Association in commentary upon the paper by Wartmann, given at that meeting, on Daltonism (see Chapter 7). The remark was reported in 1841 in *The Athenaeum* **722** 669 but not in the report issued by the British Association itself
20 Wright, W D 1952 *J. Opt. Soc. Am.* **42** 509 as quoted by Wright 1967 in *The Rays are Not Coloured* (Bristol: Adam Hilger) pp77–8

9
Further Contributions of Maxwell

MATHEMATICS AND SENSATIONS

A continuing theme throughout this history has been Young's speculation about trichromatic vision, its use by successive investigators, and its role as a model for Helmholtz and especially for Maxwell. We shall now see that Young's theory continued to serve Maxwell as his model in later investigations and that he, more than anyone, placed Young's ideas on a viable basis. Before these later works of Maxwell, however, there were two attempts to produce a theory of colour perception based on the wave theory of light, as opposed to the anatomical ideas of Young. These were by the German mathematician W J Grailich in 1854 [1] and by James Challis, the Cambridge mathematician and astronomer, in 1856. [2] Challis, especially, thought that if the wave theory were true it must be capable of explaining the facts of colour mixing. He considered the lack of such an explanation a logical gap in the mathematical framework of the wave theory. He had attempted as early as 1834 to establish an analogy between the composition of colours and the composition of vibrations, [3] and in 1856 he returned to that subject. Grailich also presumed this analogy and he worked out his ideas in somewhat greater detail than Challis. Let us see very briefly the outlines of these attempts.

In general they considered the compound variation of ether particles resulting from two different trains of waves of different wavelengths (or colours). For single vibrations (i.e. monochromatic light) the times of displacement of an ether particle on either side of equilibrium are the same. For compound vibrations (i.e. compound colours), however, the analysis shows that the times are not equal. One can assume that the momentary sensation produced by a compound vibration is the same as for a simple one when the time of displacement of the ether particle on one side of equilibrium is the same as for a simple colour. This implies that a compound vibration stimulates different colour impressions in rapid succession which are then combined in the eye into a single sensation. To achieve an analytical expression for the resulting compound wave one must assume wave trains of equal amplitude; no general analytical expression is possible for unequal amplitudes. Therefore an immediate objection to this type of theory is its severely restricted nature. Grailich's theory in particular gave a bad result, showing that combinations of red and violet would give green instead of purple. Finally, both theories beg the question as to the nature of the colour receptor mechanism, while completely ignoring the growing body of data pointing towards threefold colour perception. In 1856 Maxwell put these theories into perspective when he said of his own work:

> it proceeds upon that theory of three primary elements in the sensation of colour, which treats the investigation of the laws of visible colour as a branch of human physiology, incapable of being deduced from the laws of light itself, as set forth in physical optics. It takes advantage of the methods of optics to study vision itself; and its appeal is not to physical principles, but to our consciousness of our own sensations. [4, p245]

For Maxwell it was fruitless to search for mathematical connections between light and colour sensations. He regarded his idea for colour photography as indicating that Young's threefold receptor mechanism could be imitated by optical apparatus; and that to him implied a common anatomy in individuals, as yet undiscovered. [5, p137] Therefore he thought that research ought to be toward further knowledge of the sensations themselves through optical means and ultimately their understanding in anatomical rather than mathematical terms. It is indeed Maxwell's philosophy which has prevailed. Further attempts such as Challis's are not to be found. Colours can, of course, be understood on mathematical terms as was first shown by Grassmann and much later by Erwin Schrödinger, [6] but these theories do not consider colour mixing in terms of vibrations or wavelengths. However, a considerable insight into the anatomy of colour vision has recently begun to emerge.

ANOTHER LOOK AT MIXTURES OF BLUE AND YELLOW

In connection with his mathematical theory of colour Challis in 1856 performed a simple colour mixing 'experiment'. Placing a piece of blue ribbon on top of a yellow ribbon, he held the pair up to the light. [2, p334] He claimed the transmitted colour was green which we can understand in terms of absorption, but most amazingly he said that 'the resulting green is most decided, and is plainly due to a mixture of lights.' [2, p334] He even said that the experiment proved that 'blue and yellow light may provide a *vivid* green.' [2, p334] Mindful of Helmholtz's results but not wishing to contradict Brewster whom he esteemed, Challis concluded that there must be a physical difference between green light seen this way and that viewed through a prism. It was partly to explain this presumed difference that he constructed his mathematical theory of compound colours. He was rather severely criticised for this experiment and his conclusions by his Cambridge colleague, G G Stokes, but maintained his position nevertheless.

Challis's little 'result' shows that the paradigm of green as a primary colour for vision, as opposed to yellow, still had far to go before acceptance. Initiated by Helmholtz in 1852 and 1853, by 1856 the only other person to accept Helmholtz's results was Maxwell, and he, in any case, had come to similar conclusions independently. The difficulties of accepting the new ideas were even more evident later, in 1859, when William Pole, Professor of Civil Engineering at University College, London, still maintained yellow as a primary colour. Discussing his own colour blindness in the *Philosphical Transactions*, he said:

> It is not necessary for me to enter into the optical theory of colours generally . . . I propose to adopt the popular hypothesis that there are three primary colours, red, blue and yellow, by combinations of which all hues in nature may be supposed to be formed. [7]

This is more surprising since he knew Maxwell's paper of 1855 and even used a colour top to obtain colour equations for himself [8] yet he also compared his vision to colours in Chevreul's *Colour Atlas* by matching them with pigment mixtures which he then duplicated on the colour top. Thus he did not differentiate even yet between the two methods of mixing colours. He should indeed have entered into the optical theory of colours.

Challis and Pole are leading examples of people from whom one might have expected better things. Knowledgeable men, they nevertheless did not accept the conceptual revolution that had occurred. Therefore, Maxwell in 1855, anticipating difficulties, had aimed at avoiding controversy for using green rather than yellow by calling it a standard colour instead of a primary. A standard after all can always be changed whereas a 'primary' refers to something more basic. But Maxwell had also said that had he chosen yellow rather than green it would have been difficult to form green by any combination of blue and yellow whereas the yellow formed by his red and green was reasonably distinct. The implication was that a wider range of colour equations was possible using green rather than yellow. However, Maxwell had not pursued that question preferring to proceed with the more important matters we have already discussed.

Maxwell gave a fuller insight into that little phrase in a report to the British Association in 1856. Here he analysed his experiments from the point of view of producing mixtures of blue and yellow light. He pointed out that mixtures of yellow and blue do not make green when viewed by rapid rotation (i.e. his colour top); nor did they make green when mixed spectral colours were viewed on a screen, nor by viewing stripes of colours through a telescope out of focus at a large distance, nor by receiving the colours directly in the eye. [4, p244] The latter was an allusion to his colour box, the first version of which was originally constructed in 1852. The method of viewing colours out of focus was also due to Maxwell. [9] He even emphasised that he had made experiments with coloured powders, a kind of terminology which would look familiar to his contemporaries, but the coloured powders were actually painted on paper and used on the top so the method of employing them was really quite different from actually mixing powdered pigments. It is clear therefore from these remarks that Maxwell was quite conscious of the difficulty of gaining acceptance for a radically new idea and was attempting to introduce it in terms of the old conceptions.

The most convincing data in respect to choosing green rather than yellow came not surprisingly from the colour top experiments. Here Maxwell had explicitly tried to produce compounds from his standard colours which would match mixtures of blue and yellow. The blue, of course, would then have to be different from the ultramarine standard, and was mineral blue. The yellow (paper) was chrome yellow. Then in a systematic way Maxwell attempted to match his standard colours against blue, then against various mixtures of blue and yellow, and finally against yellow. His results are shown in Table 9.1. [4, p244] The two left-hand columns refer to the mineral blue

Table 9.1. Attempts to match mixtures of yellow and blue on the colour top with the standard colours vermilion, ultramarine and emerald green. Yellow and blue light do not yield green by additive mixing.

Given colour				Standard colours V	U	EG	Coefficient of brightness
B_8,	,	100	=	2	36	7	45
B_7	Y_1,	100	=	1	18	17	37
B_6	Y_2,	100	=	4	11	34	49
B_5	Y_3,	100	=	9	5	40	54
B_4	Y_4,	100	=	15	1	40	56
B_3	Y_5,	100	=	22	−2	44	64
B_2	Y_6,	100	=	35	−10	51	76
B_1	Y_7,	100	=	64	−19	64	109
	Y_8,	100	=	180	−27	124	277

and chrome yellow to be matched by the three standard colours, vermilion, ultramarine and emerald-green. The notation of, say, B_5Y_3, indicates a mixture of five parts blue with three of yellow as measured from the top. 100 indicates that the amount of blue and yellow occupies the full 100 divisions of the top. The values for each of the standard colours are their relative proportions required to match the blue and yellow combinations as derived from the top. Finally, the 'coefficient of brightness' is the sum of the V, U and EG parts intended simply to indicate the relative luminances. Comparing the first and second rows, we can see that the first introduction of yellow actually decreases the overall brightness, while the next several additions of yellow (with consequent decreases in blue) increase the brightness. This is to be expected. These values can be written as colour equations. Using the third row for example, we have

$$B_6 + Y_2 = V_4 + U_{11} + EG_{34}, \tag{9.1}$$

remembering that the subscripts represent only relative proportions and not actual luminance measurements from the top.

Now the important thing to see is that negative values of ultramarine are introduced when the proportion of yellow exceeds that of blue. The question is: Considering the nature of these experiments and the nature of colour equations, how do we interpret negative values? Strictly speaking, we have to write for row six:

$$B_3 + Y_5 = V_{22} - U_2 + EG_{44}, \tag{9.2}$$

but since compound colours of light are intrinsically additive, this means that a mixture of V, U, and EG cannot match the blue and yellow compound. In fact, for the required additive process we must transfer the U to the left side, and write:

$$B_3 + Y_5 + U_2 = V_{22} + EG_{44}. \tag{9.3}$$

On the top, however, transferring a colour from one side of the equation to the other means that it has changed from either an outer paper to an inner paper, or vice versa. So in fact the actual observation consisted of a match of blue, yellow and ultramarine against vermilion and emerald-green. The general conclusion, therefore, from these data is that no combination of blue and yellow light gives green, which of course, was verified qualitatively by Maxwell by other methods as well.

NEGATIVE COLOUR COEFFICIENTS

Negative values of colour in equations give further insight into the properties of primary colours. In selecting green over yellow as a primary or standard colour the question was raised: Is there one absolute set of three colours, or can any set of three be called primary? The implication from the last chapter was that any three could be used although red, green and blue seemed a better choice. We can now see why this is so, following a discussion of colour equations by Maxwell from his major paper on the theory of colour in 1860. [10, pp417–8]

Supposing there are three standard colours, x, y and z, we can, in principle, form them into a new compound colour, u, according to the equation

$$u = x + y + z, \tag{9.4}$$

where x, y and z are the luminances of each colour employed in producing u.

Conceivably, however, allowing negative values of colour, u might be formed according to the following equation:

$$u = x + y - z, \tag{9.5}$$

where we have arbitrarily chosen z negative. However, we cannot physically produce a negative colour since coloured lights mix additively. Therefore we transfer z to the other side and write

$$u + z = x + y. \tag{9.6}$$

This means that the colour $u+z$ is identical in appearance to $x+y$. On the colour top, u and z would be found on one set of papers, say the outer, and x and y would be found on the other set of papers, the inner ones. In other words, it is possible to produce a mixture of the colour u with one of the selected standards which is identical chromatically to a compound of the other two selected standards.

It is equally conceivable that the colour u might be written as

$$u = x - y - z. \qquad (9.7)$$

Then, as before, we transpose y and z and have

$$u + y + z = x. \qquad (9.8)$$

This means that a compound of two of the selected standards plus a third colour is chromatically equivalent to one of the standard colours. On the top, u, y and z would be on one set of papers, and x on the other. Interpreting negative values of colour as being transferred to the other side of the equation or to the other half of the photometer field, we must conclude that there is no ultimate, fundamental set of primary colours. With negative colour coefficients any hue can be specified by any three given colours. Maxwell expressed this conclusion by saying 'we may thus in all cases find the relation between any three colours and a fourth, and exhibit this relation in a form capable of experimental verification.' [10, pp417–8]

Why then did Maxwell originally choose vermilion, ultramarine and emerald-green (i.e. red, blue and green) as standards? In his great paper of 1855 he had explained this choice simply by saying that it was 'determined solely by the power of forming the requisite variety of [colour] combinations.' [5, p127] The answer was really that these colours produced a larger variety of combinations with *positive* coefficients than any other set of three. Considering the new concepts and methods Maxwell wished to present, employing negative colour coefficients would have been a needless complication. This idea was saved for later. Even so, we can see that had he chosen yellow rather than green even then as a standard colour he could have written that paper using negative values of yellow in some colour equations. In fact, it will be seen shortly that he actually did follow such a procedure without saying so.

THE COLOUR BOX

A difficulty with the colour top lay in its somewhat indirect operation. To produce compound colours, light had first to reflect off the various pigmented surfaces, which achieved their hues by reflecting a larger intensity of one colour than any other. So the actual selection of each single colour was a subtractive one: before producing additive colour mixtures, which of course the top did give there was first a subtractive selection process. There was also, as we have seen, a certain subjective interaction between the operator and observer which could subtly affect the measurements. What was really needed was a direct quantitative superposition of coloured lights. More specifically, since colour vision ultimately results from spectral colours, Maxwell required a means of superimposing the spectral colours in the eye. Considering the nature of his experiments, Maxwell needed a method of superimposing at least three different spectral colours in a way which would: (i) allow photometric comparison of their joint effect with a fourth colour, and (ii) allow for luminance measurements of each colour in order to form colour equations.

Maxwell's first attempt in this direction has already been noted. While on vacation from Cambridge, he mixed spectral colours in August 1852, performing experiments similar to Helmholtz's. [10, p415] [5, p144] However, this instrument was not intended to compare the compound colours photometrically with a fourth colour having 'no constant light for a term of comparison.' [10, p415] His next colour box, as Maxwell called it, was constructed in 1855 exhibiting colour combinations side by side. [10, p415] But that process, as noted earlier, requires a photometric match of two variables, hue and luminance, which complicates observation. This version, which anyway was too large and bulky even to keep steady, [11] was given up in favour of a smaller version (five feet long) which was finally perfected in 1858, providing for comparison of spectral mixtures with white (see Figure 9.2). Meanwhile another much smaller colour box (two feet long) was constructed in July 1856 and exhibited at the 1856 meeting of the British Association. [10, p415] [12] Employing folded optics it passed a beam of light through a prism twice. Maxwell's drawing of this instrument in a letter to Stokes is shown in Figure 9.1. [13, p5] [14] Like the 1852 model it did not provide for colour matching, being intended only to exhibit spectral colour combinations. It was a kind of 'proof of principle' device. This portable apparatus was finally perfected for colour matching between December 1859 and February 1860. [15] Considering the difficulties of finding suitable subjects, especially colour blind subjects, Maxwell hoped to use this small, portable colour box with larger numbers of observers than he could with the bulkier model. Nevertheless the measurements reported in the first half of his 1860 paper came from the revised 1855 model but true to form Maxwell

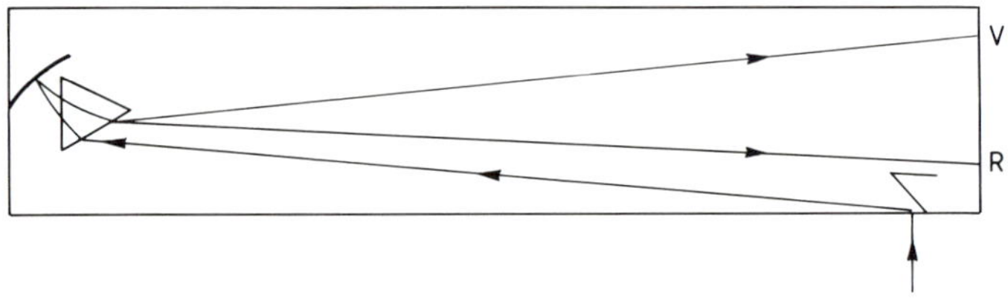

Figure 9.1 Early form of Maxwell's portable colour box as described in a letter to Stokes in 1857. [13, p5][14] It was exhibited at the meeting of the British Association of 1856. It can combine three spectral colours but cannot compare the result with white.

brought the portable model to his colour blind subject in Aberdeen to obtain data for the second half of that paper.

It is interesting to note that the colour box originated when Maxwell was also investigating geometrical optics. Letters to Stokes and Thomson in the period (1855–1859) describing the box also discussed proofs later incorporated into a paper on the laws of optical instruments. [16] His mutual interests in optics and the eye had already merged in 1854 when he derived the properties of a very wide angle lens as a result of contemplating the crystalline lens of a fish. [17] Almost certainly the colour box derived from Maxwell's dual interests in optics and colour vision at that time. In any case considering his difficulties with its earlier versions together with his other activities in physics we are not surprised that he chose to publish his results with the top first. The top, in spite of the objections raised, was a simple means of achieving reliable, repeatable results. Finally, Maxwell built yet another version of the colour box in 1862 in which the spectrum was exhibited with great purity. [18] With this version he produced some 400 colour equations none of which unfortunately were ever published. He did however use it in 1870 to examine colour vision at different points across the retina, a most modern study. [19]

Maxwell's colour box as perfected in 1858 is shown in Figure 9.1. [10, plate VI] There, two wooden tubes join at an angle of 100°. Tube AK is about five feet long, seven inches wide and four inches high while section KN is about two feet long, five inches wide and four inches high. BD is a partition.

Figure 9.2 Maxwell's colour box as finally perfected.

The inside is painted black and there is a narrow, vertical slit at E with three similar slits at end AB. Above the angle a lid can be opened to adjust the optical parts. L is a convex lens, P and P′ are equilateral prisms and a mirror of black glass is at M. Figure 9.3 is a detailed drawing of the optical parts.

The three slits at AB are shown in Figure 9.4. Sliders X, Y, and Z are attached to six moveable knife edges forming adjustable slits, which can isolate any three portions of the spectrum formed when light is admitted at

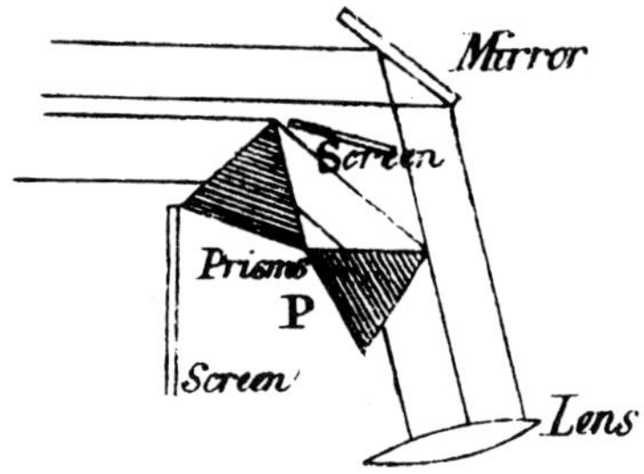

Figure 9.3 Arrangement of prisms, mirror and lens in the colour box.

slit E. A scale along which the knife edges move, calibrated in twentieths of an inch, determines the position of the slit. The actual slit width is measured by inserting a metal wedge into the slit and reading off a scale on the wedge. Calibrations of the wedge were in 200ths of an inch.

To understand the operation of the colour box suppose white light enters the instrument at slit E. Then it passes through the lens and is refracted by the two prisms at P. These prisms produce a spectrum at AB, but only the portions passing through slits X, Y, and Z will be visible. Suppose that red light is visible at slit X. By the principle of reversibility in optics the path of any ray may be reversed or an object and image may be interchanged. Therefore if white light is incident at X, only red light is visible at E. For the given prisms, slit adjustments, etc, all other light will be refracted too much or too little to reach E. An observer at E then sees the prism P illuminated uniformly in red light of exactly the wavelength allowed to exit at X when white light is admitted to E.

The same arguments apply to slits Y and Z. At Y, for instance, suppose we saw green when white light was incident at E. By the principle of reversibility, white light incident at Y will show the prisms uniformly illuminated in green when the eye is at E Therefore if slit X is also open and white light is admitted through X and Y, the observer at E sees a compound of red and green *light*. The luminance of each colour depends on the width of the slit and the brightness of the incident light. Clearly slit Z allows a third spectral colour to be compounded with the other two.

Of fundamental importance is the opening BC. White light passes through

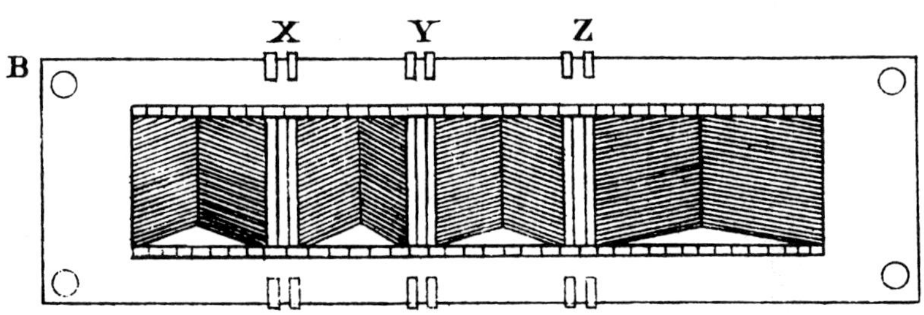

Figure 9.4 Slit arrangement at end AB of the colour box for obtaining primary colours.

here to the black mirror at M, reflecting through the lens to E. It passes close to the edge of the prisms but not, however, through them. Therefore, for white light incident at AC, the observer at E sees through the lens a field of two contiguous portions of light separated by the edge of the prism. The left-hand part consists of the colour compounded of light passing through the slits. The right-hand part is white light from the mirror. By adjusting the slits, these two fields can be made equivalent in luminance. Then, as with the top, colour equations can be formed. The instrument is then a photometer which compares the sum of three luminances in one part of the field with a single (white) luminance in the other.

Maxwell set up the box in a 'moderately lighted' room. The source, of course, was sunlight which, however, could not be introduced directly at A as it would be too bright. Outside the window was a large board covered with white paper uniformly illuminated by the sun. Then pointing AC at the board, sunlight reflected from the white paper entered the slits. This gave equal illumination to the slits and BC and also assured that any change in brightness, say due to a cloud, affected both halves of the photometer field equally. Thus a match made under cloud would be closely correct when the sun reappeared and require little readjustment. However, all measurements were made under bright sunlight, so with this arrangement he did not have to wait when the sun was behind a cloud to begin the next observations.

Thorough experimenter that he was, Maxwell also determined whether during his experiments any parts of the instrument might have moved slightly. This he did by turning the whole apparatus around and admitting light at E. Viewing at AB he located Fraunhofer's lines D and E in the slits and determined their positions on his scale before and after the measurements.

THE COLOUR BOX AS A SPECTROMETER

The colour box was ultimately conceived for 'the determination of the positions of the colours of the spectrum upon Newton's diagram, from actual observations of the mixtures of those colours.' [10, p420] This presumed that a plot of the spectral wavelengths forms a curve although not necessarily a circle as Newton had speculated. The goal was therefore to find the shape of that curve. This being so it was necessary to know the wavelengths of the spectral colours employed in the colour equations. Therefore, it was necessary to calibrate the scale at AB in terms of wavelengths which turned the colour box into a spectrometer.

To accomplish this Maxwell employed the interference bands in a thin layer of air between two pieces of glass with plane surfaces. The glasses were held apart with two parallel strips of gold leaf to make the air film approximately uniform. Referring to Figure 9.5, a ray of light of nearly perpendicular incidence on the right-hand glass experiences an internal reflection as shown. The transmitted portion of the ray is reflected from the

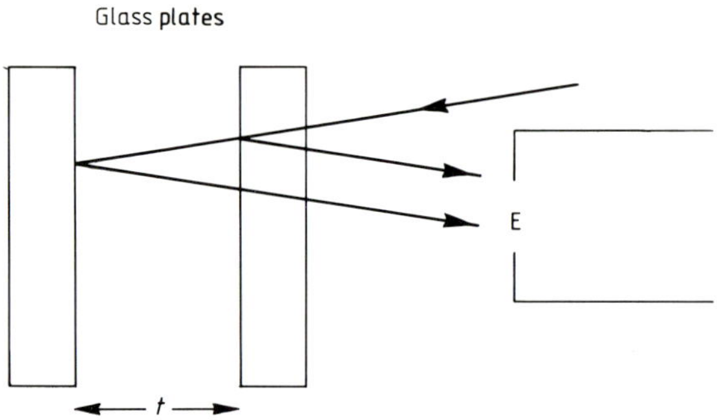

Figure 9.5 Interference in a thin film of air of thickness t between two glass plates. The plates were placed before the eyepiece, E, of the colour box and the interference pattern observed through the slits at the other end.

upper surface of the left-hand glass where it also experiences a phase change of half a wavelength. Therefore the air film creates conditions for interference between the two reflected rays. Because of the phase change of the second ray, the condition for destructive interference with monochromatic light is that the path difference (or retardation) between the rays be a whole number of wavelengths of the light, $m\lambda$. However, if t is the thickness of the air film, the path difference is also very nearly equal to $2t$. Therefore the condition for dark interference fringes is $m\lambda = 2t$. Using white light, there will be as many dark bands as there are colours whose wavelengths fulfil this condition. For similar reasons there will also be a series of constructive fringes in different colours. Maxwell placed this arrangement in the sun in front of the eye-slit at E and observed the interference bands (transmitted by the optical system) through the slits at AB. Here he viewed a series of dark fringes and also a series of coloured fringes going from red at one end of the scale to violet at the other. The question then was to find a formula to determine the wavelengths corresponding to the dark bands, which he did in terms of two well known Fraunhofer lines, D and F, that were also visible at AB. We shall not enter into the details of this process except to note that in evaluating a constant expressed in terms of the Fraunhofer lines Maxwell omitted a decimal point which then made all his calibrated wavelengths too large by a factor of 100! Table 9.2 gives Maxwell's results with the decimal point correctly placed. [20]

There were 22 bands visible for which the scale could be calibrated. Of these Maxwell chose 16 equally spaced and determined their wavelengths and apparent colours (see Table 9.2) where an extra column has been added converting the wavelength units to nanometres. [21] Since the experiments were aimed at determining colour matches in terms of three standard colours, Maxwell chose the colours at scale positions (24), (44) and (68). He expressed his reasons for this as follows:

I chose these points because they are well separated from each other on

Table 9.2. Maxwell's calibration of the scale on his colour box and his names for the colours. His values were given in millionths of a Paris inch. An extra column has been added converting his values to nanometers. In the original, the values were 100 times too large and have been corrected here.

Scale	Wavelength (millionths of a Paris inch)	Wavelength (nm)	Name of colour
20	24.50	663.2	Red
24	23.28	630.2	Scarlet
28	22.40	606.4	Orange
32	21.54	583.1	Yellow
36	20.78	562.5	Yellow-green
40	20.13	544.9	Green
44	19.51	525.1	Green
48	18.79	508.6	Bluish green
52	18.46	499.7	Blue-green
56	17.97	486.4	Greenish blue
60	17.55	475.1	Blue
64	17.21	465.9	Blue
68	16.88	456.9	Blue
72	16.60	449.4	Indigo
76	16.30	441.2	Indigo
80	16.04	434.2	Indigo

> the scale, and because the colour of the spectrum at these points does not appear to the eye to vary very rapidly, either in hue or brightness, in passing from one point to another. Hence a small error of position will not make so serious an alteration of colour at these points, as if we had taken them at places of rapid variation; and we may regard the amount of the illumination produced by the light entering through the slits in these positions as sensibly proportional to the breadth of the slits. [10, p425]

This can also be regarded as a 'sensible' statement in favour of green over yellow without antagonising the still unconverted. In the results to follow, slit X corresponds to position (24), Y to (44) and Z to (68) with the centre of the slit at the scale division.

FIRST RESULTS

In operation end AB of the box was turned towards the paper covered board illuminated by sunlight. The operator sat at AB to manipulate the slits while the observer sat at E, shaded from bright light. To make white from the three standards, the operator set the centre of the slits X, Y and Z to their respective positions and adjusted their width until the observer saw 'the prism illuminated with pure white light of the same intensity with that reflected from mirror M.' [10, p426] To accomplish this there must have been some communication between the operator and observer. For example, the observer may have required more red, or less green for which the operator would have to adjust the width of the slits accordingly. Finally if the variable field was white but darker or lighter than the constant field the operator had to suitably adjust all three slits. An initial result was given by Maxwell on observations made by himself:

$$18.5(24) + 27(44) + 37(68) = W^\star. \tag{9.9}$$

The equation means that on the day observed the width of slit X was 18.5 units, while its centre was on the scale division (24); slit Y was 27 units wide with its centre at scale division (44), slit Z was 37 units wide at scale division (68), and the white compounded from the three colours was the same luminance as the constant white, W*. The asterisk denotes the equation as a standard equation made with the three standard colours for referring other observations to white.

In all his measurements, Maxwell made colour matches to white only to avoid the problem of matching two variables, luminance and hue. As mentioned earlier he had in fact tried such matches with the first photometer version of the colour box but without success. By matching in terms of white the observational problem was reduced to comparing luminance only which allowed better judgements by the observers. But we shall see presently that there was also an important analytical reason for this procedure dealing with the standard equation.

Returning now to Maxwell's observation he then changed the position of slit X from (24) to (28) and again made a match with white. That was:

$$16(28) + 21(44) + 37(68) = W, \tag{9.10}$$

where there is no asterisk for the white since this is not made with the standard colours. If one subtracts equation (9.10) from equation (9.9) remembering that the numbers in brackets mark only slit position (hue) and not magnitude, the result is:

$$18.5(24) - 16(28) + 6(44) = 0, \tag{9.11}$$

or

$$16(28) = 18.5(24) + 6(44). \tag{9.12}$$

Thus Maxwell showed that the new colour (28) can be expressed as a compound of two standard colours (24) and (44). By combining any one colour with two standard colours, Maxwell showed that it was possible to produce a white equal in brightness to the constant white. For this, 'the red and yellow colours from (20) to (32) must be combined with green and blue, the green from (36) to (52) with red and blue, and the blue from (56) to (80) with red and green.' [10, p427] A typical series of observations made by his wife (identified as K—for Katherine) and with himself as operator gave the results in Table 9.3. The standard equation using the standard colours is marked with an asterisk. Since all other colour equations were to be

Table 9.3. Colour matches to white with spectral colours by Maxwell's wife, observer (K). The whites marked with an asterisk are the standard equation using the standard colours.

X		Y		Z	
18½(24)	+	32½(44)	+	32 (68)	= W★
17½(24)	+	32½(44)	+	63 (80)	= W
18 (24)	+	32½(44)	+	35 (72)	= W
19 (24)	+	32 (44)	+	31½(68)	= W★
19 (24)	+	30½(44)	+	35 (64)	= W
20 (24)	+	23 (44)	+	39 (60)	= W
21 (24)	+	14 (44)	+	58 (56)	= W
22 (24)	+	62 (52)	+	11 (68)	= W
22 (24)	+	42 (48)	+	29½(68)	= W
19 (24)	+	31½(44)	+	33 (68)	= W★
16 (24)	+	28 (40)	+	32½(68)	= W
6 (24)	+	27 (36)	+	32½(68)	= W
23 (32)	+	11½(44)	+	32½(68)	= W
17 (28)	+	26 (44)	+	32½(68)	= W
20 (24)	+	33½(44)	+	32½(68)	= W★
46 (20)	+	33 (44)	+	30 (68)	= W

compared as above its observation was repeated often to ensure its accuracy. This also ensured that conditions had not changed between sets of observations. There were altogether 20 observations of this equation (for K) from which Maxwell drew some interesting conclusions.

OBSERVATIONAL ACCURACY IN DISTINGUISHING COLOURS

The mean of all these observations gave for the standard equation

$$18.6(24) + 31.4(44) + 30.5(6) = W^{\star}. \tag{9.13}$$

By comparing these values with mean errors in the standard equation it became possible to judge the relative accuracy in observations of colour and brightness. Table 9.4 shows different arrangements of the errors, [10, p429] where the letters R, G, B are the standard colours (24), (44) and (68)

Table 9.4. Table of mean deviations of observational errors (MD) in the standard equation by observer (K).

MD of standard colour	MD for differences of errors of two standard colours	MD for sums of errors of two standard colours	RMS (or standard) deviation of two standard colours
(R) = 0.54	(G − B) = 0.99	(G + B) = 2.31	$\sqrt{G^2 + B^2} = 1.67$
(G) = 1.22	(B − R) = 0.85	(B + R) = 1.59	$\sqrt{B^2 + R^2} = 1.26$
(B) = 1.15	(R − G) = 0.86	(R + G) = 1.57	$\sqrt{R^2 + G^2} = 1.33$

respectively. It is immediately clear that red is the most accurately observed colour.

Now if observational errors of hues were all independent, then the deviations shown in columns 2 and 3 would be equal to the rms value of column 4. However, the deviations of column 2 are consistently smaller than the rms value while those of column 3 are consistently larger. Hence observational errors of the standard colours (or any other set) are not independent. Maxwell concluded that positive errors committed in one colour 'coincide more often with positive than with negative errors in another colour.' [10, 429–30] In other words, overreading in one colour tends to produce overreading in another colour which would not be so if the errors were independent.

From the work of Grassmann, Helmholtz and Maxwell it is apparent that the apparent hue of a compound colour depends on the ratio of the components. But the brightness of a compound colour depends on the *sum* of the components. The error analysis indicates that the difference between two colours is more accurately observed than their sum; and this implies that variations in the hue of a colour are more easily seen than variations in brightness. As Maxwell said 'the eye appears to be a more accurate judge of the identity of colour of the two parts of the field [i.e. of the colour box] than of their equal illumination.' [10, p430] This conclusion does not contradict, however, the statement that luminance judgements are best made for white, for we may consider white a colour in the sense of a physiological sensation. It means only that matchs made in terms of 'hues' other than white will have larger observational errors.

FURTHER RESULTS AND THEIR ANALYSES

The ultimate purpose of these experiments, as said before, was to plot the spectral colours on a curve by the centre of gravity method. For this, data like that of Table 9.3 were needed. One set of observations was not considered sufficient, however, especially in terms of the observational errors Maxwell had explained. Thus his wife as observer actually made four such series of observations and the mean of these were then used for analysis (see Table 9.5). [10, p430]

These equations are simply made as before by using two standard colours and substituting a new colour for the third. What is required is to reduce these data so that each particular colour can be expressed in terms of the standard colours. For this the standard equation is used to eliminate white by the method described on page 196. For example, taking the last equation in Table 9.5 we can write

$$18.6(24) + 31.4(44) + 30.5(68) = W^{\star} \qquad (9.14)$$

$$18.3(24) + 33.2(44) + 63.7(80) = W. \qquad (9.15)$$

Table 9.5. Average of four sets of colour matches to white by Maxwell's wife. These data after reduction with the standard equation to the form of Table 9.6 were then used to plot the locus of the spectral colours and the spectral mixture curves.

44.3 (20) +	31.0 (44) +	27.7 (68)	= W
16.1 (28) +	25.6 (44) +	30.6 (68)	= W
22.0 (32) +	12.1 (44) +	30.6 (68)	= W
6.4 (24) +	25.2 (36) +	31.3 (68)	= W
15.3 (24) +	26.0 (40) +	30.7 (68)	= W
19.8 (24) +	35.0 (46) +	30.2 (68)	= W
21.2 (24) +	41.4 (48) +	27.0 (68)	= W
22.0 (24) +	62.0 (52) +	13.0 (68)	= W
21.7 (24) +	10.4 (44) +	61.7 (56)	= W
20.5 (24) +	23.7 (44) +	40.5 (60)	= W
19.7 (24) +	30.3 (44) +	33.7 (64)	= W
18.0 (24) +	31.2 (44) +	32.3 (72)	= W
17.5 (24) +	30.7 (44) +	44.0 (76)	= W
18.3 (24) +	33.2 (44) +	63.7 (80)	= W

Subtracting equation (9.15) from equation (9.14), remembering that the whites are of equal luminance, we have

$$0.3(24) - 1.8(44) + 30.5(68) - 63.7(80) = 0.$$

Transposing the colour at (80) gives in terms of the standards

$$63.7(80) = 0.3(24) - 1.8(44) + 30.5(68). \qquad (9.16)$$

Hence, using the standard equation we can express the wavelength at (80) in terms of the standard colours. The negative coefficient of colour (44) means that such a match cannot be physically produced. That colour would have to be transferred to the left side of the equation or to the other side of the photometer field to make an actual match. This was not possible with Maxwell's apparatus. Nevertheless, as Maxwell had shown in 1856 the coefficient does allow us to specify colour (80) implicitly in terms of the three standard colours.

Repeating this process with the other colour equations gives the results in Table 9.6 [10, p430] where equation (9.16) appears on the last line. Each spectral colour has now been expressed in terms of the spectral standards, where the symbols of the standards are placed at the top of each column. Maxwell was now in position to draw a curve of the spectral colour in terms of the given colours. Note that colour (46) is not in the calibration list.

The method of plotting the wavelengths is essentially the same as that of the paper of 1855. An equilateral triangle was the coordinate system with the standard colours at the vertices. The position of each wavelength was found by locating its centre of gravity from the reduced data of Table 9.6. Let us plot, for example, the positions of (20) and (56). The equation for (20) is:

$$44.3(20) = 18.6(24) + 0.4(44) + 2.8(68).$$

Referring to Figure 9.6 we imagine weights of 18.6, 0.4 and 2.8 units placed

Table 9.6. Reduced form of the spectral colour matches by Maxwell's wife.

	(24)	(44)	(68)
44.3 (20) =	18.6	+ 0.4	+ 2.8
16.1 (28) =	18.6	+ 5.8	− 0.1
22.0 (32) =	18.6	+19.3	− 0.1
25.2 (36) =	12.2	+31.4	− 0.8
26.0 (40) =	3.3	+31.4	− 0.2
35.0 (46) =	− 1.2	+31.4	+ 0.3
41.4 (48) =	− 2.6	+31.4	+ 3.5
62.0 (52) =	− 3.4	+31.4	+17.5
61.7 (56) =	− 3.1	+21.0	+30.5
40.5 (60) =	− 1.9	+ 7.7	+30.5
33.7 (64) =	− 1.1	+ 1.1	+30.5
32.3 (72) =	+ 0.6	+ 0.2	+30.5
44.0 (76) =	+ 1.1	+ 0.7	+30.5
63.7 (80) =	+ 0.3	− 1.8	+30.5

at R, G and B of the triangle respectively. Let P be the centre of gravity between (24) and (68). We can then write: 18.6 RP = 2.8 PB, which locates P when the length of the side is known. Next we divide the line PG at a point Q, and write (18.6 + 2.8)PQ = 0.4QG. Again knowing the length of PG, we locate Q which, by definition, is the point corresponding to the colour (20).

For (56) the equation is:

$$61.7(56) = -3.1(24) + 21.0(44) + 30.5(68).$$

We can write this instead as

$$61.7(56) + 3.1(24) = 21.0(44) + 30.5(68).$$

The centre of gravity between (44) and (68) is point D found by the process above. The centre of gravity of (56) plus (24) must also lie at D. Since the position of (24) is known, the position of (56) must lie on a line from R through D on the other side of D. This procedure was followed in the last chapter to locate the position of pale chrome. Therefore draw the line RD

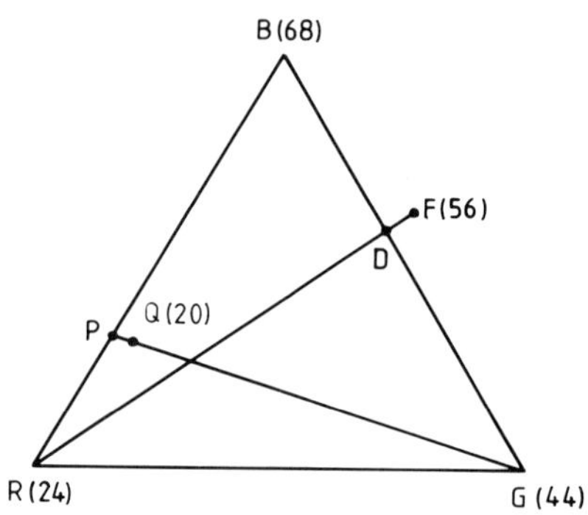

Figure 9.6 Locating colours (20) and (56), red and blue-green respectively, in terms of the standard colours from observations by (K).

and extend it to F. We write now: 3.1RD = 61.7DF, or DF = (3.1/61.7)RD, which locates (56) since RD can be measured.

Maxwell's complete plot for all the colours in Table 9.6 is shown in Figure 9.7, [10, Plate VI] where the vertices labelled A, B and C are red, green and blue respectively. Connecting the plotted points shows that the spectrum locus forms two sides of a triangle 'with doubtful fragments of a third side.' [10, p431] We may also consider that if any three colours lie in a straight line, the middle colour is a compound of the other two and not necessarily desaturated as had been previously thought. Therefore it is possible to extrapolate from this diagram the three spectral colours from which all others are compounded. They must lie at the vertices of the *plotted* locus, rather than being the standard colours actually used for the matching experiments.

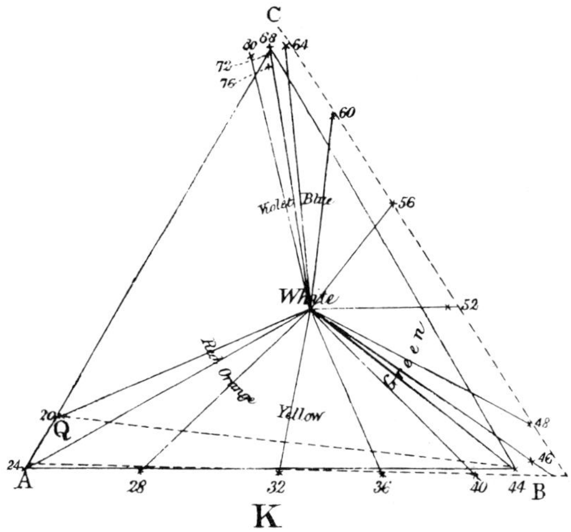

Figure 9.7 Locus of the spectral colours as observed by (K).

Maxwell estimated R = (24), a scarlet; G = ($46\frac{2}{3}$), and B = ($64\frac{1}{2}$). Since all natural colours are ultimately compounded from spectral colours, they may also be compounded from these three. Indeed, for the first time, Maxwell referred to these as primary colours and he had done nothing less here than actually measure the primary colours whose existence until now had been wholely conjecture. It would also imply that the colour (24) is equivalent to point D of Figure 8.7. However, Maxwell did not claim to have actually determined this.

These results are strictly true only for the one observer. For generality they would have to be repeated for many people. Although Maxwell was quite aware of this he was not able to investigate many subjects. Still to confirm these methods and ideas, he made similar observations on himself, presumably with his wife as operator. The procedure followed was exactly as before and the results were similar but certainly not identical. For himself the standard equation became

$$18.1(24) + 27.5(44) + 37(68) = W^{\star}, \qquad (9.17)$$

while a plot of his colour matches gave the locus of Figure 9.8. [10, Plate VI]

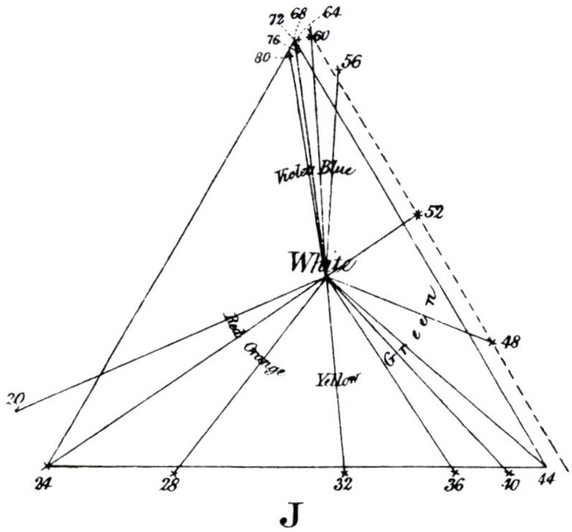

Figure 9.8 Locus of the spectral colours as observed by Maxwell.

For Maxwell the colours (32), (36) and (44) were further to the right which means that he saw more green in these colours than his wife. Secondly the colours (48), (52), (56) and (60) are higher up the side of the triangle than for his wife, which means that they appeared more blue to him than to her. Such results were confirmed experimentally. A match made by (K) at colours (32), (36) and (40), did not match for Maxwell, one side being too green. Similarly for colours (48) to (60), matches made by (K) appeared too blue for Maxwell. In reverse, a match on the green side of the triangle for Maxwell appeared too red to (K) while the other matches appeared too green. Therefore, although both sets of eyes were normal there were noticeable differences between them in the perception of colours. From measurements on himself, three primaries cannot be extrapolated and it is significant to note the absence of an attempt to do so from Maxwell's paper. In terms of the question concerning whether people see colours alike, the answer is obviously yes; however, they are certainly not seen identically, there being real measurable differences. Clearly, Maxwell's wife had better colour perception than he did.

Pursuing this discovery, Maxwell made colour matches to white on himself in two ways. By looking directly at a colour match he employed the fovea, that portion of the retina of most acute vision. Then he averted his vision slightly which caused the prism side of the photometer field to appear greener than the mirror side. Therefore his colour vision some distance away from the centre of vision was more sensitive to the blue-green region of the spectrum than was the fovea itself. Thus he made an important discovery; colour vision across the retina is not uniform so that observed results from this method of colour equations actually depend on the size of the field of view. (A difference in colour sensitivity across the retina had actually been noted earlier by the Czech physiologist Evangelista Purkinje but Maxwell was unaware of this.)

However, his partial blue-green insensibility was restricted to those wavelengths near slit position (52). Nor was it a colour blindness due to a lacking sensation in the sense of Young. For then the match made by direct (foveal) vision would remain matched by the averted (extra-foveal) vision. A missing sensation in the sense of Young would presumably operate uniformly across the retina which was certainly not the case here. Here the green sensation was reduced in the centre of vision only, which Maxwell attributed to the yellow pigment covering the fovea. It was this yellow spot about which he had disputed with Brewster in 1850 and whose polarising effect he had demonstrated later in relation to Haidinger's brushes in 1856. [22] He had realised even then that the retina did not sense all colours equally but the actual effects he had not been able to measure. Specifically he had noticed an elongated dark spot running up and down in the blue portion of the spectrum, following the eye's motion and never leaving the blue. He had then attributed the effect to the yellow spot inferring that it was due to absorption.

Now in 1860 he had specific data. He attributed his weakened blue-green sensitivity to specific absorption by the yellow spot in that portion of the spectrum. Hence he was certainly not colour blind, but colours ultimately reaching the retina were altered in proportion by absorption in the yellow spot. Significantly his wife had no yellow spot. It was this discovery which stimulated Maxwell to his final colour box of 1862 with which he studied more closely colour vision across the retina. Detailed results from these researches however were never published, intimations of them being given only in letters and at the British Association meeting of 1870. In respect to the yellow spot he said in a letter of 1870 that

> I can exhibit the yellow spot to all who have it,—and all have except Colonel Strange, F.R.S., my late father-in-law, and my wife,—whether they be Negroes, Jews, Parsees, Armenians, Russians, Italians, Germans, Frenchmen, Poles, etc. [23]

COLOURS OUTSIDE THE TRIANGLE

It has been said before that the interior of the triangle of colours (and its perimeter) define all the colours compoundable from those primaries. Yet Maxwell in 1855 and in 1860 plotted several colours outside the triangle, which so far have been given no interpretation.

Compare the data of Table 9.6 with the locus of spectral colours, Figure 9.7. It is seen that every colour in the table with a negative coefficient falls outside the triangle of standard colours. Thus strictly speaking we cannot compound this colour from the primaries. But, as we saw with colour (56), we *can* plot its position. In that case the negative colour was transposed to the other side of the equation meaning that a photometric match could be obtained, in principle, between the colours on the left and right sides of the equation. Then we plotted the position of (56). Colours outside the triangle mean then that they are specified by negative coefficients. They cannot be compounded from the primaries themselves but they can still be defined in terms of those primaries and the operational method of colour matching.

In the 1855 announcement of his work, Maxwell showed how to plot the position of pale chrome which fell outside the triangle. That procedure, after normalisation, was exactly the same as for colour (56). In fact in his 1855 plot of the colours (Figure 8.6) all the colours were outside the triangle. Maxwell was then using negative colour coefficients without saying so explicitly. That idea was reserved for a later announcement.

There is one important difference between Maxwell's plotting method of 1855 and that given here. Using coloured papers, a normalisation procedure was necessary to plot the locus of each colour in order to refer all the colour equations to a constant luminosity white. For that purpose the colour top equation

$$0.37\ \mathrm{V} + 0.27\ \mathrm{U} + 0.36\ \mathrm{EG} = 0.28\ \mathrm{SW} + 0.72\ \mathrm{Bk}$$

played the role of a standard equation just as equation (9.13) did here. But in these researches no normalisation procedure had been necessary. The colour points were plotted directly from the data after reduction with the standard equation assuming all whites were of equal luminance, since they were all observed under the same conditions from the same constant source. Normalisation was now accomplished at the time of observation rather than arithmetically. In essence, the methods of 1855 and 1860 were the same except the new data were far more accurate and referred to real wavelengths. They also applied to a much smaller field of view than those with the top. Significantly Maxwell in 1855 did not attempt fitting a curve to his points (Figure 8.6) but only indicated it was roughly a circle. Now he joined the points and found that the curve is in fact not a circle nor even a closed curve. It is roughly similar to Helmholtz's hypothetical curve of 1855. The curve can be considered closed however between red and violet through purple transitions, recognising that this represents a visual extrapolation from the data.

THE CHROMATICITY DIAGRAM

Maxwell's work must be considered as the origin of the modern chromaticity diagram. This is a curve whose points are the loci of the spectral colours and whose values are derived from the colour mixing data. It represents in different form the data of colour mixing and from it one can obtain additional colour information as well. [24] However, the actual numerical data are ultimately based on the selection of primaries and other experimenters could have (and have) selected different wavelengths than Maxwell. Indeed in this science the primaries are always somewhat arbitrary and even vary from one such photometer to another. Maxwell's primaries were arbitrary in that he selected them within certain ranges of red, green and blue. But Grassmann's third law means that light mixtures can be treated independently of their spectral compositions; two matching colours can have quite different spectral content. Therefore, the selection of primaries ultimately needn't matter. The data can be plotted with respect to one set of primaries or with respect to any other. Considering any set of primaries as a set of coordinates, the data of one system can be transformed into another system by a transformation of axes where we need only the desired transformation equations between one set of colour standards and another. While this was a later development, Maxwell himself did briefly consider this very idea without pursuing it. Suffice it to say here that spectrum loci are now based on standards adopted in 1931 and are frequently plotted in rectangular coordinates as in Figure 9.9. While the shape of Maxwell's curve changes, essentially the same information is still conveyed. In Figure 9.9 Maxwell's data is plotted for (K), his wife, together with combined data of two later investigators and the 1931 standards. [25]

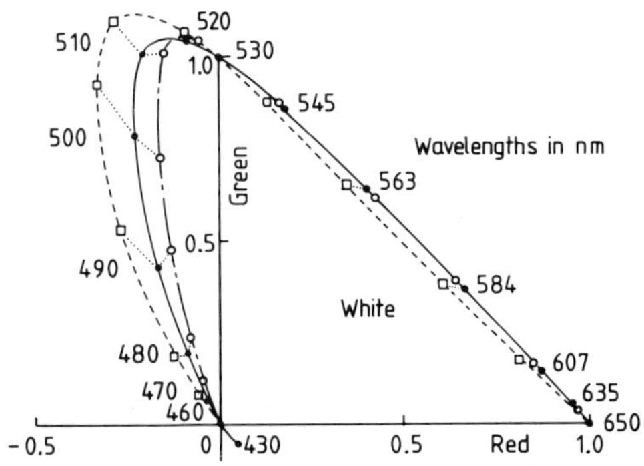

Figure 9.9 Spectrum loci as observed by different experimenters. Open circles, Maxwell 1860 (observer K); open squares, Konig, Abney 1903, 1913 (recalculated by Weaver); and full circles, CIE standard observer 1931 (Wright, Guild).

Now in Maxwell's colorimeter the slit width measured luminance, but changing the widths also changed the band of wavelengths actually admitted at that setting so none of the colours were really monochromatic. Also we cannot be sure of the equal luminances of all the whites assumed by Maxwell's method. Considering these difficulties, the team of Maxwell and Maxwell did rather well.

These results can also be understood from another point of view, briefly mentioned at the end of Chapter 6. Following Grassmann's vector ideas, which Maxwell certainly had done, a plot of any colour must be three-dimensional. As in Figure 9.10, three axes in space must represent the three standard colours. Then, knowing the luminance of each component of a given colour, a point is plotted representing that colour. A line from the origin to the point is a vector representing the colour and the vector sum of the components yields the length of the vector which is proportional to the

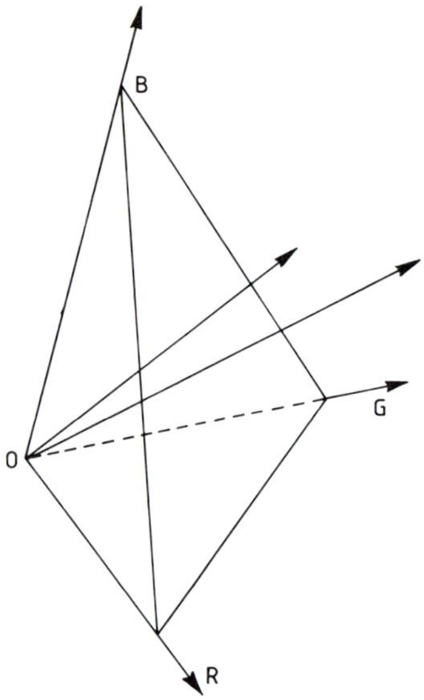

Figure 9.10 Three-dimensional plot of colours with respect to red, green and blue primaries. Colours are vectors with origin at O. The triangle cuts off equal lengths on the axes intersecting all colours of equal luminance. This is equivalent to a projection from the top of Lambert's pyramid onto the base yielding the chromaticity diagram. See Figure 6.8, where the projection was made for a cone instead of a tetrahedron.

light luminance. If unit lengths of the three primaries are laid off on the axes and the points joined, an equilateral triangle is formed which is a plane passing through the plot of colours giving a surface of constant illumination. We can say that the colour triangle and the chromaticity diagram associated with it is a projection onto a plane of the colours plotted in three dimensions. It is equivalent to projecting the colours from the top of Lambert's pyramid to the base, where in Lambert's terms the base of his pyramid would represent unit luminosity. Any other plane parallel to the base of Mayer's and Lambert's pyramids also represents a colour triangle of uniform, although different, illumination.

The location of any colour C in the triangle can be represented by the equation

$$C = rR + gG + bB, \tag{9.18}$$

where R, G and B represent the standard colours and r, g and b are now called the chromaticity coordinates. From this equation it is clear that $r + g + b = 1$ on the unit plane. In the triangle each coordinate represents the perpendicular distance from the side of the triangle to the colour point and their actual values are derived from the colour matching data. However, the location of a colour in the triangle does not of itself give any information about the actual luminance of the colour. The chromaticity coordinates simply define the proportions of the standard colours required to match that hue. Thus for any other triangle parallel to that shown in Figure 9.10 the proportions of r, g and b for C remain constant and its location in that triangle is the same as in the unit triangle. Regardless of the actual luminance value assigned to the triangle of Figure 9.10, colours of either high or low luminance are still located at the same point in the triangle. It follows that Newton's colour circle and the triangles of Mayer and Lambert were early forms of the chromaticity diagram, qualitative in nature. They represented only colour matches with positive coefficients (colours inside the triangle) and did not allow for negative coefficients (colours outside the triangle). That the colour triangle could be given this interpretation was not realised, of course, by any of the colour classifiers, nor was it recognised then by Helmholtz. It was Maxwell's work that yielded an experimental justification and theoretical understanding of those systems. Maxwell's curves were the first chromaticity curves, and it is still essentially by Maxwell's methods that data for the modern curves are obtained.

Finally, it is not actually necessary to plot the spectrum loci in terms of a triangle. Since $r + g + b = 1$ any two values specify the third and a triangle actually introduces a redundant coordinate. For this reason Figure 9.9 is plotted in rectangular coordinates. If, however, a triangle is used, an equilateral triangle, taken as an article of faith since Mayer's time, is also not necessary. Mayer we know used this figure as a reduced form of Newton's circle. It has the clear advantage of treating the selected primaries in an easy, symmetrical fashion, which is certainly the reason for its widespread popularity. But the primary axes in Figure 9.10 can be oriented with respect to each other in any way desired so the shape of the triangle formed from them is arbitrary. In principle any triangle can be used to plot the chromaticity diagram. In the same sense as coordinate transformations, the shape of the curve changes but the information conveyed by it remains constant. Regardless of the shape of the triangle, it still represents a projection from the origin of three-dimensional axes onto a plane.

THE FIRST LUMINOSITY CURVE

Maxwell's observations are subject to yet more analysis. Considering Table 9.6 we can take each equation and divide through by the coefficient of the left-hand side. Thus the first equation there is

$$44.3(20) = 18.6(24) + 0.4(44) + 2.8(68),$$

and dividing by 44.3 gives

$$(20) = 0.42(24) + 0.009(44) + 0.063(68).$$

This puts the equation into the form of equation (9.18) expressing unit luminance for the wavelength in terms of the three standard colours. Doing this for each colour in the table, it becomes possible to plot the relative luminosity of each standard colour as a function of wavelength. There will then be three distribution curves each of which represents a physiological response mechanism in the eye. Maxwell did precisely this and his results for his wife and himself are shown in Figures 9.11 and 9.12 respectively. [10, Plate VI] The curve marked R is the luminosity distribution of the standard red (24), G is for the standard green (44), and B is for the standard blue (88).

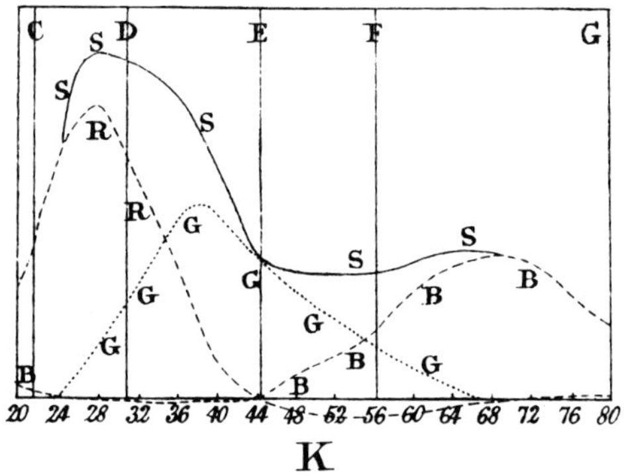

Figure 9.11 Spectral mixture curves for observer (K).

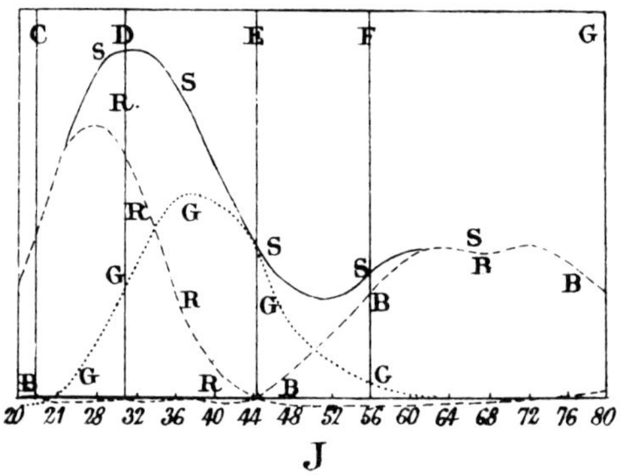

Figure 9.12 Spectral mixture curves for Maxwell.

Curve S is the sum of the ordinates of the separate curves. The vertical lines C, D, E, F and G are positions of various Fraunhofer lines. Each curve, being an individual physiological response, shows how each primary stimulates a perceived sensation at each part of the spectrum.

Let us compare these curves with a similar set proposed by Helmholtz. In 1860, shortly before Maxwell's paper appeared, Helmholtz published part of his definitive work *Physiological Optics*. [26] Provisionally accepting Young's concepts, he had traced hypothetically, for discussion only, the response curves corresponding to Young's processes (Figure 9.13). His curves simply expressed Young's original idea that each receptor would have a maximum response in one region of the spectrum and diminished response throughout the rest of the spectrum. Curve 1 is for the red-sensitive 'fibres', curve 2 for

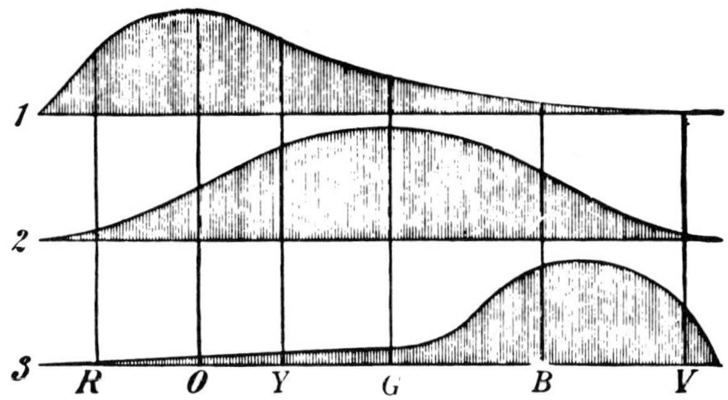

Figure 9.13 Helmholtz's hypothetical curves for Young's three colour receptors. Compare with Figures 9.11, 9.12 and 9.14.

the green and curve 3 for the violet, which Helmholtz, following Young, chose rather than blue.

Maxwell had now succeeded in doing nothing less than demonstrating experimentally the responses about which Young had so long ago speculated. That Helmholtz's interpretive curves and Maxwell's actual curves are different is not suprising. On the contrary there is a remarkable similarity between them which adds additional credence to the trichromatic hypothesis. With Young the idea had been purely speculative. But the evolution of events after him had been towards increasing the indirect evidence favouring this idea. The three-dimensional colour classifications, Grassmann's linguistic evidence, Maxwell's geometrical demonstration of the equivalence of hue, shade and tint with three standard colours, and the colour blind data all weighed in favour of Young's theory. Perhaps the best evidence, but still indirect, was Maxwell's indication of the point D in Figure 8.7. Yet what was ultimately needed to verify Young's theory was direct evidence of three sensations. It is precisely this that Maxwell had provided. The speculative nature of the hypothesis had been converted by Helmholtz into a prediction that such response curves should be obtainable, and Maxwell, independently of Helmholtz, recognising the same requirement, produced data which confirmed that prediction. In this work, Maxwell (as opposed to Helmholtz) transformed Young's speculation into a meaningful empirical statement.

The results above are equally as important as the chromaticity diagram. Like that curve the response curves of Figures 9.11 and 9.12 illustrate the colour matching data but they also give some insight into the physiological nature of colour perception. These curves represent essential facts which 'must provide the foundation of any adequate theory of colour vision.' [27] A great deal of work after Maxwell, continuing into recent times, has been devoted to achieving improved data for spectral mixtures. Moreover, the old objection to Young concerning lack of anatomical evidence for the receptors now took on a new meaning. That too has been a challenge for the twentieth century.

As noted earlier, the equal luminance white used in this programme was equivalent to the normalisation process used in the earlier paper. There the standard papers had unit normalisation factors in principle, or simply unit brightness. In this new approach, the luminosity curves of Figures 9.11 and 9.12 are 'intended to indicate the intensities of the three standard colours at different points of the spectrum.' [10, p433] Given the curves one can read the proportions of red, green and blue primaries required to compound any colour. However, it follows from equation (9.18) that the luminosities of the standard colours themselves should be unity which they clearly are not in Maxwell's curves. The reasons for this deal with his measurement of luminance in units of slit widths (200ths of an inch) and the probable variations from equal luminance of the whites for different sets of observations. Appropriate correction factors can, however, be estimated for Maxwell's data to convert them to modern standards and units. This was performed by D B Judd in 1961 whose results are shown in Figure 9.14. [28]

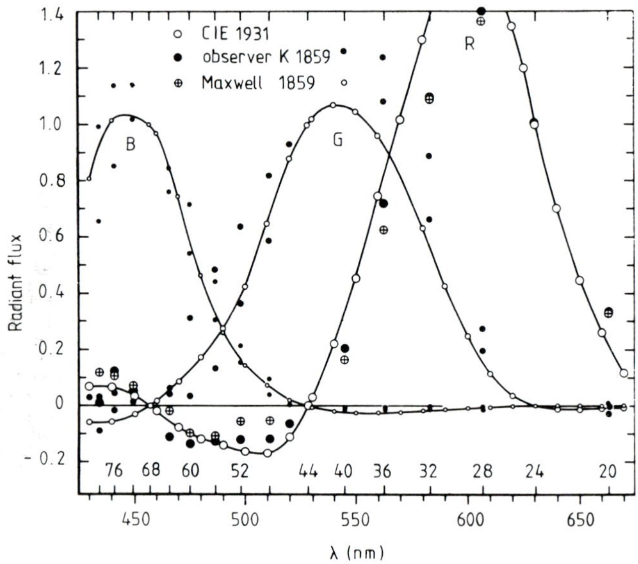

Figure 9.14 Maxwell's data normalised to conform with modern observational standards and units and compared with the 1931 standard CIE observer using Maxwell's primaries. The correction was made by Dean B Judd of the US Bureau of Standards.

The curves plotted with open circles are those of the 1931 CIE standard observer employing Maxwell's spectral primaries. (The slight differences in values between Maxwell's primaries used by Judd and those given in [21] are because Judd did not notice Maxwell's arithmetic error.) Then Maxwell's observations for (K) and (J) are plotted against these for comparison. Once more we can see that the Maxwell team gave a rather good account of itself—and this on a first try. Maxwell not only devised correct procedures for determining spectral luminosity curves, but his own observations compare quite well with modern values.

The results discussed so far were sent by Maxwell to G G Stokes in

October, 1859. [13, pp11–12] Maxwell was thinking of presenting the results to the Royal Society saying at the end of the letter

> If you think any of these results proper aliment for the Royal Society pray tell me how I may administer it in due form, as I do not purpose to be in London.
>
> The subject is one about which we have had conversation before, so that I have not thought it necessary to do more than indicate new results as you know by what method they must be obtained and how they are expressed . . . [13, p14]

Stokes' reply was obviously positive. The paper, communicated to the Royal Society by Stokes, who was then its Secretary, formed Maxwell's first publication in the *Philosophical Transactions*. In fact he was nominated on the basis of this paper to be Bakerian Lecturer for 1860. However, he was unable to be in London at a suitable time to deliver it so the lecture was ultimately delivered by someone else on a different topic. (Maxwell ultimately did deliver a Bakerian lecture in 1866 on the measurement of the coefficient of viscosity of gases.) To a certain extent the paper was incomplete for the observations above were produced by normal persons. Their ultimate purpose had been to produce a spectrum locus like that of Figure 9.7 and luminosity curves like those of Figures 9.11 and 9.12. Maxwell proposed to Stokes, however, that the paper ultimately read in March 1860 be considered 'Part I of a more complete one'. [13, p14] He wished to have observations with a colour blind subject in order to form colour blind equations to add confirmation to the theory of three-fold colour vision. [13, p14] For this he intended to use the improved, portable colour box but was delayed because his subject was ill. However, between the end of March and mid-April, 1860, Maxwell's subject, a former student of his at Marishal College, Aberdeen, was sufficiently recovered to produce satisfactory observations. Maxwell wrote to Stokes on 23 April describing some of these results, [13, pp18–9] and a postscript to his original paper was received by the Royal Society on May 8. Maxwell's reason for awaiting the good health of his colour blind student was crucially important, for obtaining consistently good observations was not as easy as this history might suggest. We had a hint of this difficulty in the last chapter where we saw that Maxwell rejected observations by two colour blind observers as not being sufficiently accurate or consistent. Now Maxwell said quite plainly with respect to observations with the colour box by a normal observer that:

> It is difficult at first to get the observer to believe that the compound light can ever be so adjusted as to appear to his eyes identical with the white light in contact with it. He has to learn what adjustments are necessary to produce the requisite alteration under all circumstances, and he must never be satisfied till the two parts of the field are identical in colour and illumination. To do this thoroughly, implies not merely good eyes, but a power of judging as to the exact nature of the

difference between two very pale and nearly identical tints, whether they differ in the amount of red, green, or blue, or in brightness of illumination. [10, pp415–6]

If this was true for normal sighted persons it was equally true for colour blind observers in that they too needed a feeling for the requirements of these measurements. This necessity was apparently fully appreciated by Maxwell's student. In explaining to Stokes why he had waited for his students to recover Maxwell wrote 'it is not very easy to find colour blind persons who know what is required to make good observations.' [13, p20] The difficulties of finding competent observers explain partially why Maxwell investigated so few people, but it also underlines the important point that he was particularly fortunate in having found this student for his investigations.

COLOUR BLIND OBSERVATIONS

Before looking into Maxwell's colour blind data let us examine his new colorimeter (Figure 9.15). [10, Plate VII] As in the other apparatus, light enters through three adjustable slits X, Y and Z. It then passes through prisms P and P′ (with refracting angles of 45°), to a concave silvered mirror S.

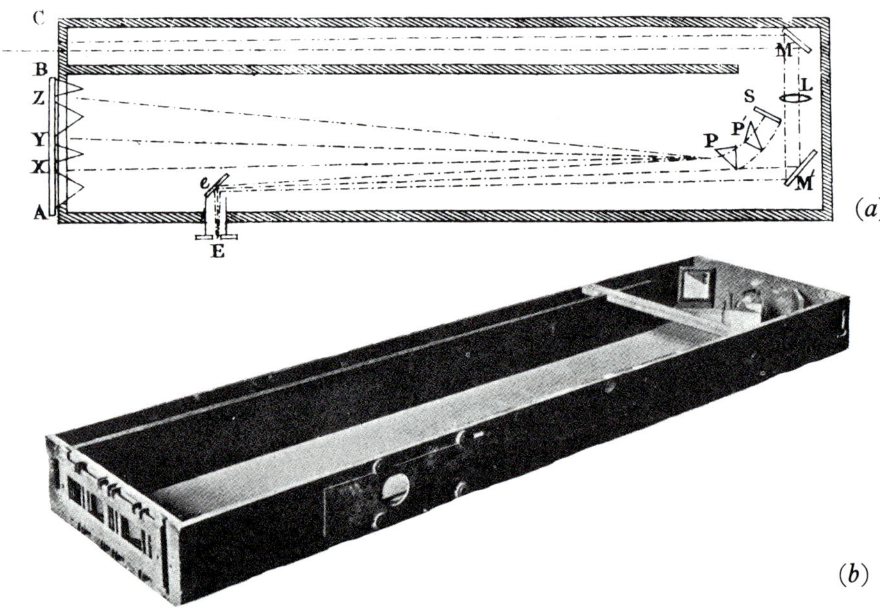

Figure 9.15 Maxwell's portable colour box. (*a*) Schematic; (*b*) actual (from Gunther, R T 1937 *Early Science in Cambridge*, courtesy of Oxford University Press).

It then passes back through the prisms to a small plane mirror at e and through the observer's slit at E. The observer then sees a colour at E compounded of three spectral wavelengths.

At BC white light passes through the channel to plane mirror M, through the lens L, and reflects again from mirror M′ to the mirror at e. Notice that light from M′ passes close by the edge of prism P. The observer therefore sees two contiguous fields, one of colour and the other of white light. In this early example of folded optics, the improvement comes from reflecting the light back through the prisms which eliminates the limb KN in the previous device. More importantly, it enables the other limb (AK in the first box) to be considerably shortened while simultaneously giving greater angular dispersion to the spectrum. The apparatus produced a spectrum 3.6 inches long between Fraunhofer's lines A and H but was only $3\frac{1}{2}$ feet long and was therefore quite portable. Maxwell with typical wit even claimed in May 1860 in a letter to G G Stokes that it was 'warrented to be safe when turned upside down and moderately jolted.' [13, p16]

Maxwell had already shown in his earlier work that vision for so-called red blind individuals is a function of two variables. For these protonopes similar to Dalton (or Seebeck's Class II) the spectrum from orange through green appears as a 'yellow' of varying intensities. Their vision of the red portion of the spectrum is also so curtailed that in Maxwell's words 'there is not enough in the whole red division of the spectrum . . . to make up the standard white.' [10, p437] Finally the blue-violet section of the spectrum appears to them mainly blue. Maxwell found that for such persons, the green at Fraunhofer's line E appears a 'good yellow' while the blue two-thirds the distance from Fraunhofer's line F toward green appeared a 'good blue'. [10, p438] By this he meant that the colours appeared saturated and pure although how he determined this is not clear. However, by this time there was considerable knowledge of red blind vision so we may accept this statement as by now constituting an application of general knowledge. Given these facts, Maxwell chose the two spectral positions above as the standard colours for his two-fold or 'dichromic' observations. [10, p438] [29] On the scale of his new instrument green had the position (88) and blue was at (68). For comparison with *normal* vision, a red standard was also required to make white, which position became (104), although the red blind observer could not see this.

For Maxwell's colour blind student an average of ten readings to make white gave the standard equation for him as

$$33.7(88) + 33.1(68) = W^{\star}. \tag{9.19}$$

Once again an error analysis showed that the mean of the sum of the errors was larger than the mean of the difference. Therefore, colour blind individuals as well as normal persons detect distinctions in hue more easily than differences in luminance. In terms of a photometric match, observation of equality of hue (regardless of how interpreted by the colour blind observer) will be more dependable than those of equality of brightness for both classes of people.

As a normal subject for comparison, Maxwell reasonably enough used himself. His new standard equation from a mean of six observations became

$$22.6(104) + 26(88) + 37.4(68) = W^{**}. \quad (9.20)$$

We can now rewrite the colour blind equation (9.19) as

$$D + 33.7(88) + 33.0(68) = W, \quad (9.21)$$

where D, as in Chapter 8, represents the missing sensation. Then we can combine equations (9.20) and (9.21) by eliminating W assuming, of course, that the luminances are equal. This gives for D

$$D = 22.6(104) - 7.7(88) + 4.3(68).$$

Therefore, we can define D in terms of the three standard colours. We can not, however, define the optical wavelength of D, something which was only done much later. [30] Still, it is possible now to plot D according to Maxwell's methods and if the spectrum locus for this colorimeter were plotted as before, the optical wavelength of D could be estimated against the curve. However, the data here for both colour blind and normal observations, are each for one individual only. Considering the differences in normal individuals shown in the first part of his paper, another normal person would produce a slightly different spectral locus, which would necessarily yield a different value for D. Many observers would be required from which to achieve an average standard equation for comparison with colour blind individuals. Moreover, the point D will vary among red blind persons just as there are variations among normal observers. Hence, any estimate of the wavelength of D is necessarily an average value. (It appears from modern work that D lies at about 700 nm.) The necessity for more observers was recognised by Maxwell who, however, did not have time in his busy schedule for finding them. It is startling nevertheless to realise that Maxwell had it within his grasp to estimate the optical wavelength of D but simply lacked the opportunity to collect the necessary data.

Returning to Maxwell's colour blind subject, he was required to perform colour matches in the same manner as normal observers. Then using his own standard equation, these observations were reduced as before to a table where each particular wavelength is expressed in terms of two standard colours. These are shown in Table 9.7. [10, p439] The first things to notice here are the first and last entries in the left-hand column. The top of the scale reads (99.2+). This is the red end of the spectrum but from Maxwell's own normal standard equation the red scale positions extended to at least (104). This indicates that from the position (99.2) to the extreme red end of the spectrum all colour sensations for this person were identical. The bottom scale position is (57−), where the minus sign means positions toward the violet end of the spectrum. This position is in the blue. Similarly, therefore, all wavelengths beyond this position to the extreme violet produced the same colour sensation.

As before the data can be converted into a spectrum locus but here Maxwell used a different method of plotting. Since the data are functions of two variables only, a triangle is not needed to plot the chromaticity diagram.

Table 9.7. Colour matches performed by Maxwell's colour blind student. Reds on the scale of the instrument extended from about (96) on the orange side to at least (104) which Maxwell described as appearing a 'full red' to himself. This observer needed no reds to perform his colour matches.

	(88)	(68)
(99.2+) =	33.7	1.9
31.3 (96) =	33.7	2.1
28 (92) =	33.7	1.4
33.7 (88) =	33.7	0
54.7 (84) =	33.7	6.1
71 (82) =	33.7	15.1
99 (80) =	33.7	33.1
70 (78) =	15.7	33.1
56 (76) =	5.7	33.1
36 (72) =	− 0.3	33.1
33.1 (68) =	0	33.1
40 (64) =	0.2	33.1
55.5 (60) =	1.7	33.1
(57−) =	− 0.3	33.1

Rectangular, two-dimensional coordinates suffice. The axes consist of the two colours seen by the observer, his 'yellow' or (88) and his blue or (68). We recall that what he meant by 'yellow' appeared as green to a normal observer, so the axes are simply green and blue. Then one divides through the data of Table 9.7 by the coefficient in the left-hand column and that gives the amount of each standard colour to plot for a given scale position. Drawing a curve through each point, the result was that of Figure 9.16. [10, Plate VII]

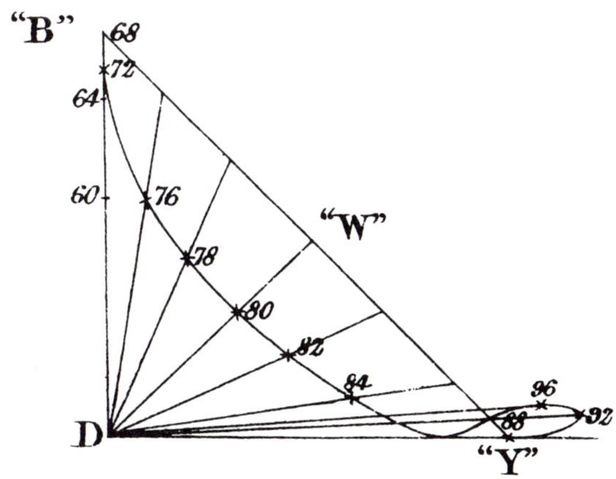

Figure 9.16 Locus of the spectral colours according to Maxwell's colour blind observer.

Maxwell took the position of point D 'as the origin of coordinates', [10, p440] without further comment. We can, however, get some insight into this choice based on the previous discussion of plotting colours in three or two dimensions. In three dimensions, colours can be plotted in rectangular coordinates as shown in Figure 9.10. In a projection of the three-dimensional plot along the red axis onto the B–G plane, the R-axis degenerates into the origin giving a two-dimensional plot. But in Maxwell's colour blind subject it was exactly the red sensation that was missing. So D in Maxwell's plot is equivalent to the red axis suppressed from the three-dimensional plot by projection onto the B–G plane. In other words, the missing axis is the analogue of the missing sensation. This is the modern point of view which Maxwell does not seem to have explicitly realised. There was, however, a hint of this understanding for he did say that 'the triangle of colours is reduced, in the case of dichromic vision, to a straight line "B" "Y".' [10, p440] Then by drawing a line from D to each spectral colour and producing it to BY, one could find by the ratios the proportion of each standard colour appearing in the given spectral colour.

Finally these data can be used to plot luminosity curves as before. Having divided by the coefficient of each colour the data have been reduced to unit luminosity. Then the relative luminosity of each standard colour is plotted against the scale position (or wavelength) giving the spectral distribution for that standard. Maxwell's plot is shown in Figure 9.17. [10, Plate VII] As before various Fraunhofer lines are shown in terms of the scale. The curve on the left is the green ('yellow') response, that on the right the blue response. In this case the resultant or overall response is simple since there is little overlap of the separate curves. The broken line in the middle represents the sum of the two curves, so the overall luminosity curve is simply the outline of the green curve to just past G, then the broken line, and then the blue curve.

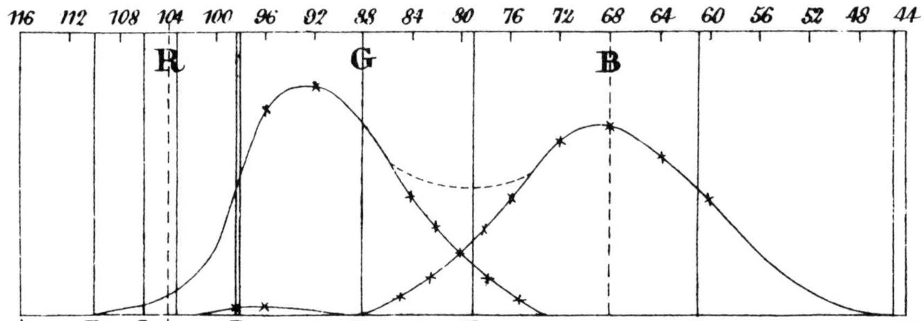

Figure 9.17 Spectral mixture curves for Maxwell's colour blind observer.

At this point Maxwell had achieved his stated goal of determining the spectrum locus by Newton's method for normal vision and had compared such results with a similar plot for abnormal vision. Following the intuitive hint of Newton's regarding the centre of gravity of colours, the qualitative work of Helmholtz and the explicit theory of Grassmann, Maxwell had converted the science of colour perception from an almost non-experimental one into a well developed empirical science with its own methods and means of analysis. Modern researches in colour perception can be dated from Maxwell.

These researches illustrate a relatively little known side of Maxwell's scientific career. He is best known, of course, as the theoretical physicist 'par excellence'. Besides his famous experiment on the viscosity of gases, he was also the first director of the Cavendish Laboratories where important physical measurements were made during his time, but he was not so much personally involved with this work. The work described in the last two chapters shows that besides his unsurpassed theoretical abilities, he was a first rate experimental physicist as well. It is a mark of the difficulties to be overcome in this field that a man of Maxwell's stature was stimulated by it and saw a challenge to overcome. Like so much else that this genius touched, he left the science transformed.

THE REVIVAL OF YOUNG'S THEORY OF VISION: A REAPPRAISAL

It is usually said that Young's theory of colour vision was generally ignored until it was resurrected by Helmholtz. For example, P J Bouma, a physicist, wrote in 1936 that 'the study of the phenomena of colour vision has led to the enunciation of two different sets of theories . . . the first, due originally to Young and Helmholtz, assumes that three sensations can take place in the eye.' [31] The juxtaposition of the names here implies that little or nothing took place in the period between the two men. Alexander Wood, also a physicist, wrote in his biography of Young (1954) that

> Young's work on colour sensation attracted no immediate attention. It was not until Helmholtz rediscovered it, nearly fifty years later, that it became effective. He adopted Young's theory with some modification, and it has since come to be known as the Young–Helmholtz theory. [32]

More recently R Steven Turner writing on Helmholtz in 1972 said that 'in his paper of 1852 Helmholtz . . . revived Thomas Young's forgotten theory of colour vision.' [33] Other works are hardly different, always associating Young and Helmholtz so it seems there is a leap from Young to Helmholtz with nothing in between.

This interpretation depends partly on Helmholtz's use of Young's ideas in his 1852 paper on colour mixing, but there he had not accepted Young's hypothesis by any means. He had even thought that possibly five primary colours were necessary for colour vision, an error which was pointed out by Grassmann. Had he recognised the possibility of negative colour coefficients he could have supported the three-receptor theory then even with his doubtful observations, but that idea, of course, was yet to come. In his follow-up work of 1853 Helmholtz did not even mention Young but concentrated mainly on determining spectral complementary pairs and their wavelengths. Much more influential for this truncated history was undoubtedly his *Physiological Optics* where Helmholtz stated in 1860:

> Young's theory of the colour sensations, like so much else that this

> marvellous investigator achieved in advance of his time, remained unnoticed, until the author [i.e. Helmholtz] and MAXWELL again directed attention to it. [26, p163]

The general opinion of a fifty year silence on Young's concepts must thus be traced to Helmholtz himself in his *Physiological Optics*, which was so widely read and studied. From Helmholtz to recent times each writer on colour perception has simply obtained his 'historical' facts from the previous writers and perpetuated the idea.

However, the study of colour blindness in Chapter 7 shows that Young's ideas were actually fairly well known, especially in Britain. Brewster in his commentary on Butter's article in 1822 referred explicitly to Young's concepts of vision. Harvey in 1826 alluded to them while Brewster again discussed them in public in 1826. John Herschel definitely accepted the main outlines of these ideas in his letter to Dalton in 1833 and did so earlier in his *Treatise on Light* in 1827. There, however, he was less explicit than in his letter to Dalton although he did use the term 'dichromic' to indicate two-fold colour vision. And Herschel's *Treatise* was translated into French and German making it accessible in Europe. In the medical profession too, Mackenzie mentioned Young's ideas in his book on diseases of the eye. In Britain, at least, Young's concepts were not only known and discussed but were even regarded with some favour. Unfortunately, after 1831 Brewster's triple spectrum competed with Young's theory and tended to confuse the issues. For Brewster believed in the objective reality of the three colours in the spectrum and therefore in their reality for visual purposes also, whereas Young had speculated only on nerve sensations in the eye regardless of the reality of the colours. Young could therefore easily change his position on the specific colours involved in vision as he did following Wollaston's observations. But Brewster dogmatically maintained his position of red, yellow and blue since these colours uniquely supported his presumptions. Accordingly, Brewster's spectrum as a theory of vision was an obstacle to further understanding of colour vision. It was, of course, Helmholtz who removed this obstacle.

On the Continent also Young's idea received a hearing and limited acceptance. Seebeck discussed the various hypotheses regarding colour blindness in 1837 while not subscribing to any. Wartmann, three years later, thoroughly reviewed all the hypotheses on colour vision mentioning Young explicitly; he showed that Young's hypothesis was the only one without objection. His memoir was translated into English in 1846 giving it a wider circulation and his opinion was repeated in another memoir of 1848. Finally Wartmann in turn influenced further favourable opinion towards Young in Britain especially through Wilson who had not read Helmholtz's paper of 1852. Thus instead of a blanket ignorance of Young's theory until the time of Helmholtz there was on the contrary a continued knowledge and reference to it from 1822 onward. Far from being ignored, Young's ideas of colour perception were given due consideration, if not acceptance, by people interested in colour blindness. Therefore the opinion that Young's ideas were ignored for fifty years, initiated by Helmholtz, cannot be supported.

What was the case is that Young's theory did initially receive a poor reception largely as result of Lord Henry Brougham's devastating attacks. Concentrating on Young's theory of light, Brougham did not even mention his theory of colour vision, but since that idea evolved from Young's wave considerations, so vehemently denounced by Brougham, anyone following his arguments and rejecting Young's wave concepts of light would perforce reject his theory of colour vision also. However, with Fresnel's first memoir on light in 1816 and especially with his subsequent memoirs, Young was in effect 'rehabilitated'. After that, Young's wave theory papers were highly regarded so those who were involved in any colour perception researches would naturally know of his idea. Is it possible then that Helmholtz was initially ignorant of Young's theory? And if so, why did he not know of it?

As to the first question Helmholtz himself freely admitted that at the time he was working on the nature of vision he was indeed ignorant of Young's work. In a popular lecture in 1868, Helmholtz recalled his earlier experiences when he had been searching for an explanation for colour perception. He said,

> I long toiled at the task, without getting any nearer my object, until I at last discovered that a wonderfully simple solution had been discovered at the beginning of this century, and had been in print ever since for anyone to read who chose. This solution was found and published by the same Thomas Young who first showed the right method of arriving at the interpretation of Egyptian hieroglyphics. [34]

But why Helmholtz should not have known of Young's hypothesis is less clear. It is hard to believe that a person of Helmholtz's far ranging knowledge should have been ignorant of ongoing researches in colour blindness, a field so closely allied to his own. Yet the fact that he read Newton's *Opticks* in its Latin translation indicates that he was not then familiar with the English literature on the subject where so much material lay. [35] Still there was sufficient work in German, especially that of Seebeck, to acquaint him with Young.

It seems that Helmholtz was looking for an overall concept of colour vision while the people involved with colour blindness were looking more narrowly for concepts of defective colour vision. These were subtly different points of view. In this respect, it is significant that Maxwell first placed the three-receptor theory on a firm empirical basis; only then, within that context, did he demonstrate red blindness as an absent response as Young had speculated. Moreover, those writers on colour blindness spoke of Young's work in rather restricted, often naive, terms with little to offer by way of detail. They hardly wrote in depth on what Young had said nor did they generate empirical data to support him. That we have seen was Maxwell's role. In short, they simply considered Young's theory as one possible explanation for their observations but otherwise made little use of it. Therefore, even if Helmholtz were aware of these researches he may well have viewed them in a different light. In any case, he certainly believed that he had rediscovered Young's theory independently and was therefore justified in making his statement of 1860. Even then, however, Young's hypothesis for him was still only a conjecture, simply the best at the time. While he did propose three response curves in reaction to that theory, it was Maxwell who actually produced them experimentally. Helmholtz's book being published in its first edition in 1860 was written, of course, before Maxwell's latest work. In a supplement to his book in 1868, Helmholtz described Maxwell's 1860 work giving a slight commentary on it. He did not, however, reproduce Maxwell's curves nor indicate any improvements of his own in Maxwell's results. We must conclude that even for Helmholtz, Young's theory was still one mainly for discussion, albeit with far greater insight to its possibilities than his predecessors.

As opposed to Helmholtz, Maxwell early accepted Young's ideas. He was certainly influenced in this by his contacts with Forbes and Wilson and his reading of Herschel's *Treatise*. [36] In this sense he lay within the British tradition of favourable response to Young. Writing to Wilson in January 1855 Maxwell said:

> The theory which I adopt assumes the existence of three elementary sensations, by the combination of which all the actual sensations of colour are produced . . . the important part of the theory is not that three elements enter into our sensations of colour, but that there are only three. [37]

From the last two chapters it is clear that Maxwell had adopted this theory in his preliminary experiments in 1854 and very likely as early as 1852. From the beginning Maxwell accepted Young's trichromatic hypothesis and its implications and made it his goal to place it on an observable, empirical basis. It was he who generated observational evidence supporting Young's hypothesis; within that framework he also confirmed Young's later explanation of Dalton's colour blindness as the absence of the red receptor. He proved also that the artistic impression of three variations in colour, hue, saturation and brightness, first clarified by Grassmann, was formally equivalent to the other centuries old impression of three primary colours. Finally whereas Herschel and Helmholtz had only presumed that any three colours could be used as primaries, it was Maxwell who showed how it must be understood experimentally. His photometric matching method together with negative colour coefficients allowed any colour to be specified, in principle, in terms of any three standard colours. Thus in Maxwell's hands Young's early thoughts took on a new operational meaning.

In conclusion then, there was certainly no fifty year silence on Young's theory as Helmholtz believed, but there was certainly no progress with it either in that period. Helmholtz implicitly recognised this point in saying that 'the author and MAXWELL' directed attention to it. We should really understand Helmholtz to have meant that there was now an entirely new viewpoint on the significance of Young's hypothesis, much different from anything before it, and much of that new insight was due to Maxwell's work. Considering this, the term 'Young–Helmholtz' theory of colour perception would be better named the Young–Helmholtz–Maxwell theory.

NOTES

1 Grailich, W J 1854 Beitrag zur Theories der gemischten Farben *Sber. Akad. Wiss., Wien, Math. Nat. Kl.* **13** 14. The few remarks here are based on [26]

2 Challis, James 1956 On the theory of the composition of colours on the hypothesis of undulations *Phil. Mag.* Ser. 4 **12**

3 Challis, James 1834 *Rep. Br. Assoc.*

4 Maxwell, James Clerk 1856 On the theory of compound colours with reference to mixtures of blue and yellow light *Rep. Br. Assoc.* Reprinted 1890 in *The Scientific Papers of James Clerk Maxwell* Vol I (London: Cambridge University Press)

5 Maxwell, James Clerk 1856 Experiments on colour, as perceived by the eye, with remarks on colour blindness *Trans. R. Soc. Edinb.* **21**. Reprinted 1890 in *The Scientific Papers of James Clerk Maxwell* Vol I (London: Cambridge University Press)

6 Schrödinger, Erwin 1920 Outline of a theory of colour measurement for daylight vision *Ann. Phys., Lpz.* **63** See also 1926 Thresholds of colour differences *Müller–Pouillets Lehrbuch der Physik* 2nd edn

7 Pole, William 1859 On colour blindness *Phil. Trans. R. Soc.* 324

8 Colour tops with papers made to Maxwell's description were available from an optician named J M Bryson in Edinburgh. In his paper of 1855 Maxwell had mentioned Bryson as his supplier and it was from him that Pole obtained his top

9 Campbell, Lewis and Garnett, William 1882 *The Life of James Clerk Maxwell* (London: Macmillan) p214. Letter from J D Forbes to Maxwell, dated 16 May 1855. In discussing whether blue and yellow light make green, Forbes refers in the letter to the telescope viewing procedure as one 'you proposed'

10 Maxwell, James Clerk 1860 On the theory of compound colours, and the relation of the colours of the spectrum. Reprinted 1890 in *The Scientific Papers of James Clerk Maxwell* (London: Cambridge University Press)

11 Larmor, Joseph (ed) 1937 *Origins of Clerk Maxwell's Electric Ideas, as described in familiar letters to William Thomson* (London: Cambridge University Press) p17. In the letter of 13 September 1855 Maxwell mentions having 'great difficulties' in obtaining good measurements from this instrument. For remarks on his final perfection of this instrument see Maxwell's letter to C J Monro dated 29 April 1858 in Campbell, Lewis and Garnett, William 1882 *The Life of James Clerk Maxwell* (London: Macmillan) p309. The 1855 model was $6\frac{1}{2}$ ft long compared to 5 ft for the revised 1858 version

12 The report of the British Association meeting of 1856 does not mention Maxwell's colourbox

13 Larmor, Joseph (ed) 1907 *Memoir and Scientific Correspondence of the late Sir George Gabriel Stokes* Vol II (London: Cambridge University Press)

14 The device is also mentioned in a letter, dated 17 December 1856, to William Thomson in Larmor, Joseph (ed) 1937 *Origins of Clerk Maxwell's Electrical Ideas, as described in familiar letters to William Thomson* (London: Cambridge University Press). Here there is a hand drawn diagram; the diagram in [13] was redrafted there for publication

15 Letter to G G Stokes, 24 December 1859. 'An optician here (Aberdeen) is making me a reflecting colour apparatus with which I intend to get colour

blind equations among the prismatic colours which will determine accurately the missing colour'. [13, p14] In another letter to Stokes, dated 25 February 1868, Maxwell refers to his colour boxes and says 'I have now got a reflecting one'
16 Maxwell, James Clerk 1858 On the general laws of optical instruments *Quart. J. Pure Appl. Math.* **II**. Reprinted in 1890 in *The Scientific Papers of James Clerk Maxwell* (London: Cambridge University Press) pp271–85
17 Maxwell, James Clerk 1854 Solutions to problems *Cambr. Dubl. Math. J.* Reprinted in 1890 in *The Scientific Papers of James Clerk Maxwell* pp76–8
18 Letters from Maxwell to C J Munro, dated 18 February 1862 and 6 July 1870 in Campbell, Lewis and Garnett, William 1882 *The Life of James Clerk Maxwell* (London: Macmillan) p334 and pp346–7
19 Maxwell, James Clerk 1870 On colour vision at different points of the retina. Reprinted in 1890 in *The Scientific Papers of James Clerk Maxwell* (London: Cambridge University Press) pp230–34. The original announcement was at the British Association meeting of 1870
20 Maxwell also made a small arithmetic error in evaluating the constant. In Table 9.2 I have expressed Maxwell's own values as they should have been given but have otherwise not recalculated them. In converting the wavelengths to nm the equivalence is 1 Paris inch = 27.07 mm as given by the US National Bureau of Standards, Office of Weights and Measures. See further D B Judd [28]
21 Using the corrected value of Maxwell's constant, his primary wavelengths were 629.7 nm, 527.9 nm and 456.7 nm, not those indicated in Table 9.2
22 Maxwell, James Clerk On the unequal sensibility of the foramen centrale to light of different colours. Reprinted in 1890 in *The Scientific Papers of James Clerk Maxwell* (London: Cambridge University Press) p242. The paper was originally presented at the British Association Meeting of 1856
23 Letter from Maxwell to C J Munro, dated 6 July 1870 (see [18])
24 See, for example, Crowell, Thomas Y 1953 *The Science of Color* (New York: Optical Society of America)
25 Everett, C W F 1967 Maxwell's scientific papers *Appl. Opt.* **6** 645
26 von Helmholtz, Hermann 1860 *Physiological Optics* 1st edn (Hamburg and Leipzig) p143
27 Wright W D 1946 *Researches on Normal and Defective Colour Vision* (London: Henry Kimpton) pp143–4
28 Judd, D B 1961 Maxwell and modern colorimetry *J. Photog. Sci.* **9** 349
29 Maxwell borrowed the word 'dichromic' from Herschel. The modern equivalent term is 'dichromatic'
30 Pitt, F H G 1935 *Med. Res. Counc. Spec. Rep.* Ser. No. 200 as quoted by Wright, W D 1967 *The Rays are not Coloured* (Bristol: Adam Hilger) p77
31 Bouma, P J 1936 The perception of colour *Philips Tech. Rev.* **1** 287
32 Wood, Alexander 1954 *Thomas Young, Natural Philosopher* (London: Cambridge University Press) p112
33 Turner, R Steven 1972 Helmholtz in *Dictionary of Scientific Biography* Vol 6 (New York: Charles Scribner's Sons)
34 von Helmholtz, Hermann 1884 The recent progress in the theory of vision *Popular Lectures on Scientific Subjects* Ser 1. English translation by E Atkinson (London) pp219–20
35 In his paper of 1853 Helmholtz quoted Newton several times from the Latin version
36 Maxwell mentions using Herschel's *Treatise* in his paper of 1855
37 Maxwell, James Clerk 1855 Letter to George Wilson. Reprinted in 1890 in *The Scientific Papers of James Clerk Maxwell* (London: Cambridge University Press) p21

Bibliography

PRIMARY SOURCES

Airy, George Biddell 1847 The Astronomer Royal on Sir David Brewster's new analysis of solar light *Phil. Mag.* Ser. 3 **30**

Arago, François 1857 *Biographies of Distinguished Scientific Men* (London)

—— 1858 *Oevres Complètes de François Arago* (Paris and Leipzig)

Aristotle 1955 *Minor Works on Colour* (English translation by W S Hett) (Cambridge, Ma: Harvard University Press)

—— 1960 *On the Heavens* (English translation by W K C Guthrie) (Cambridge, Ma: Harvard University Press)

—— 1962 *Meteorlogica* (English translation by H D P Lee) (Cambridge, Ma: Harvard University Press)

The Athenaeum 1841 Report of the meeting of the British Association for 1841

Birch, Thomas 1757 *The History of the Royal Society of London* (London). Facsimile reproduction 1968 (New York: Johnson Reprint Corp)

Biot, Jean Baptiste 1815 Sur la cause de la coloration des corps *J. Mines* **38**

—— 1816 Construction d'un Colorgrade *Bull. Sci. Soc. Philomantique, Paris*

Bouguer, Pierre 1760 *Traité d'Optique sur la Gradation de la Lumière* (English translation by W E Knowles Middleton 1961) (Toronto: University of Toronto Press)

Boyle, Robert 1664 *Experiments and Considerations Touching Colours* (London). Facsimile reproduction 1964 (New York: Johnson Reprint Corp.)

—— 1688 *Some uncommon observations about vitiated sight* appended to *A Disquisition on the Final Causes of Natural Things* (London) Reprinted in 1772 *Works of Boyle* ed Thomas Birch vol V (London) pp445–52

Brewster, David 1822 Observations on vision through coloured glasses, and their application to telescopes, and to microscopes of great magnitude *Edinb. Phil. J.* **VI**

—— 1823 Description of monochromatic lamp for microscopical purposes, with remarks on the absorption of the prismatic rays by coloured media *Trans. R. Soc. Edinb.* **IX**

——1826 On the invisibility of certain colours to certain eyes *Edinb. J. Sci.* **4**
—— 1829 Account of two remarkable cases of insensibility in the eye to particular colours *Edinb. J. Sci.* **10**
—— 1831 On a new analysis of solar light *Edinb. J. Sci.* **V** Also 1834 *Trans. R. Soc. Edinb.* **XII**
—— 1833 *Letters on natural magic addressed to Sir Walter Scott* (London)
—— 1838 *Optics* (London)
—— 1847 Observations on the analysis of the spectrum by absorption *Phil. Mag.* Ser. 3 **30**
—— 1847 Sir David Brewster's reply to the Astronomer Royal on the new analysis of solar light *Phil. Mag.* Ser. 3 **30**
—— 1848 Observations on the elementary colours of the spectrum in reply to M Melloni *Phil. Mag.* Ser. 3 **32**
—— 1855 *Memoirs of the Life, Writings and Discoveries of Sir Isaac Newton* (Edinburgh). Republished 1965 (New York: Johnson Reprint Corp.)
British Association Reports 1834, 1841, 1850, 1855
Brougham, Henry 1802 Review of Young's 'Theory of light and colours' *Edinb. Rev.* **1**
—— 1803 Review of Wollaston's 'A method of examining refractive and dispersive powers by prismatic reflection' *Edinb. Rev.* **2**
—— 1803 Review of Wollaston's 'On the oblique reflection of Iceland crystal' *Edinb. Rev.* **2**
—— 1803 Review of Young's 'Account of some cases of the production of colours' *Edinb. Rev.* **2**
—— 1804–5 Review of Thomas Young's 'Experiments and calculations relative to physical optics' *Edinb. Rev.* **5**
Butter, John 1822 On a peculiar case of vision *Edinb. Phil. J.* **6**
Challis, James 1834 Theoretical explanation of some facts relating to the composition of the colours of the spectrum *Rep. Br. Assoc.*
—— 1856 A theory of the composition of colours on the hypothesis of undulations *Phil. Mag.* Ser. 3 **12**
Chevreul, M E 1859 *The Principles of Harmony and Contrast of Colours* 3rd English edn (London)
—— 1861 Exposé d'un moyen de defines et de nommes les couleurs d'après une méthode précise et expérimentale, avec l'application de ce moyen à la définition etc, des couleurs *Mem. Acad. Sci., Paris* **33**
—— 1864 *Des couleurs et de leurs Applications aux Arts Industrielles a l'Aide des Circles Chromatiques* (Paris)
Cohen, I B 1958 *Isaac Newton's Papers and Letters on Natural Philosophy* (London: Cambridge University Press)
Dalton, John 1798 Extraordinary facts relating to the vision of colours: with observations *Mem. Lit. Phil. Soc., Manchr* **5**
Dove, H W 1852 On the stereographic combination of colours, and on the influence of brightness on the relative intensity of colours *Ann. Phys., Lpz.* **83, 85** (English translation by J Tyndall 1852 *Phil. Mag.* Ser. 4 **4**)

Draper, John William 1847 On the production of light by heat *Phil. Mag.* Ser. 3 **30**

Exley, Thomas 1834 *Physical Optics: or the Phenomena of Optics Explained According to Mechanical Science* (London)

Field, George 1820 Aesthetics, or the analogy of the sensible sciences *The Pamphleteer* **17**

—— 1835 *Chromatography; or a Treatise on Colours and Pigments and their Powers in Painting* (London)

—— 1858 *Rudiments of the Painter's Art* (London)

—— 1870 *Rudiments of Colours and Colouring* Revised edn by Robert Mallet FRS (London)

Forbes, J D Correspondence and Papers of J D Forbes (St Andrews: University Library Archives)

—— 1849 Hints towards a classification of colours *Phil. Mag.* Ser. 3 **4**

—— 1855 Observations on Mr Maxwell's Paper *Edinb. New. Phil. J.*

Foucault, Léon 1853 Sur la récomposition des couleurs du spectra en teintes plates *Cosmos, Rev. Encycl. Hebd. Prog. Sci.* **II**

von Goethe, Johann Wolfgang 1840 *Goethe's Theory of Colours* (English translation with notes by George Eastlake) (London)

Grassmann, Hermann 1854 On the theory of compound colours *Ann. Phys., Lpz.* (English translation *Phil. Mag.* Ser. 4 **7**)

Harvey, George 1826 On an anomalous case of vision with regard to colours *Trans. R. Soc. Edinb.* **X**

Harris, Moses 1963 *Natural System of Colours* Facsimile reproduction (Faber Birren, private publication)

Hay, D R 1845 *A Nomenclature of Colours, Hues, Tints and Shades* (Edinburgh and London)

—— 1846 *First Principles of Symmetrical Beauty* (Edinburgh and London)

Hayter, Charles 1826 *A New Practical Treatise on the Three Primitive Colours* (London)

—— 1845 *An Introduction to Perspective Drawing and Painting* (London)

von Helmholtz, Hermann 1852 On Sir David Brewster's new analysis of solar light *Ann. Phys., Lpz.* **8** (English translation 1852 *Phil. Mag.* Ser. 4 **4**)

—— 1852 On the theory of compound colours *Ann. Phys., Lpz.* **9** (English translation *Phil. Mag.* Ser. 4 **4**)

—— 1884 *Popular Lectures on Scientific Subjects* Ser. 1 and 2 (English translation by E Atkinson with an introduction by J Tyndall)

—— 1911 *Physiological Optics* 3rd edn (Hamburg and Leipzig). English translation by J P C Southall 1924 (New York: Optical Society of America)

Herschel, John F W 1823 On the absorption of light by coloured media, and on the colours of the prismatic spectrum exhibited by certain flames *Trans. R. Soc. Edinb.* **12**

—— 1845 Article on light *Encycl. Metrop.* **IV**

Horner, J K 1817 Description d'un photomètre *Bibl. Univers. Sci. B.-Lett. Arts* **6**

Huddart, J 1777 An account of persons who could not distinguish colours *Phil. Trans.* **67**

Larmor, Joseph (ed) 1907 *Memoir and Scientific Correspondence of the Late Sir George Gabriel Stokes* (London: Cambridge University Press)

—— 1937 *Origins of Clerk Maxwell's Electric Ideas as Described in Familiar Letters to William Thomson* (London: Cambridge University Press)

LeBlon, James Christopher 1731 *Coloritto; or the Harmony of Colouring in Painting Reduced to Mechanical Practice* (London)

Leslie, John 1800 Description of an hygrometer and photometer *J. Nat. Phil. Chem. Arts* **3**

—— 1813 Description of an atmometer, and an account of some photometric, hygrometric and hygroscopical experiments *Tilloch Phil. Mag.* **XLII**

—— 1827–8 Letter from Professor Leslie to the Editor on Mr Ritchie's experiments on heat, and a new photometer *Edinb. New Phil. J.* **4**

Lichtenburg, Georg Cristoph 1775 *Opera Inedita Tobiae Mayerii* (Göttingen)

Maxwell, James Clerk Correspondence and Papers of J C Maxwell (Cambridge, University Library)

—— 1890 *The Scientific Papers of James Clerk Maxwell* (London: Cambridge University Press)

Melloni, Marcedinio 1848 Researches on the radiations of incandescent bodies, and on the elementary colours of the solar spectrum *Phil. Mag.* **32** Also in 1847 *Bibl. Univers. Sci. B.-Lett. Arts*

Moigno, L'Abbé 1850 *Répertoire d'Optique Moderne* (Paris)

Mortimer, Cromwell 1831 *An Account of Mr James Christopher LeBlon's Principles of Painting in Imitation of Painting and of Weaving Tapestry, in the same manner as Brocades* (London)

Müller, Johannes 1835–40 *Handbuch der Physiologie des Menschen für Vorlesungen* (Coblenz)

Newton, Isaac 1730 *Opticks* 4th edn (London)

—— 1959 *The Correspondence of Isaac Newton* (London: The Royal Society and Cambridge University Press)

Nicod-Delom, J S 1816 Description d'un photomètre nouveau *Bibl. Univers. Sci. B.-Lett. Arts* **1**

—— 1834 Observations sur la photomètre de M. Nicod-Delom *Bibl. Univers. Sci. B.-Lett. Arts* **55**

Nobili, Leopold 1837 Memoirs on colours in general, and particularly on a new chromatic scale deduced from mettalochromy for scientific and practical purposes *Bibl. Univers. Sci. B.-Lett. Arts* **44, 45** (English translation 1837 *Taylor's Scientific Memoirs* **1**)

Ostwald, Wilhelm 1931 *Colour Science* (London: Winsor and Newton) (English translation by J Scott Taylor)

Palmer G 1785 *Lettre sur les Moyens de Produire, la Nuit, une Lumière Pareille à celle du Jour* (Paris)

Pole, William 1857 On colour blindness *Phil. Mag.* Ser. 4 **13**

Prevost, Pièrre 1827 Sur la vision extraordinaire de J B *Bibl. Univers. Sci. B.-Lett. Arts* **35**

Priestley, Joseph 1772 *The History and Present State of Discoveries Relating to Vision, Light and Colours* (London)

Reade, Joseph 1816 *Experimental Outlines for a New Theory of Colours, Light and Vision* (London)

Ritchie, William 1825 On a new photometer, with its application to determine the relative intensities of artificial light, etc *Phil. Trans. R. Soc.* **65**

—— 1826 On a new photometer, founded on the principles of Bouguer *Trans. R. Soc. Edinb.* **X**

Rood, Ogden 1879 *Modern Chromatics* (New York)

Runge, Philip Otto 1810 *Farbenkugel, oder Konstruktion des Verhältnisses aller Mischung der Farben zu einander, und ihrer vollständiger Affinität* (Hamburg)

Schrödinger, Erwin 1920 Grundlinien einer Theorie der Farbenmetrik im Tagessehen *Ann. Phys., Lpz.* **63**

Scott, J 1779 An account of a remarkable imperfection of sight *Phil. Trans. R. Soc.* **68**

Seebeck, A 1937 Uber den bei mancher Personnen vorkommenden Manger an Farbensinn *Ann. Phys., Lpz.* **42**

Shairp, J C, Tait, P G and Reilly, A A 1873 *Life and Letters of James David Forbes* (London)

Swan, William 1849 Phenomena of vision *Trans. R. Soc. Edinb.*

—— 1870 Obituary to Sir David Brewster *Lit. Phil. Soc., St Andrews*

Syme, Patrick 1821 *Werner's Nomenclature of Colours* 2nd edn (Edinburgh)

Tait, Peter Guthrie 1898 Obituary notice to James Clerk Maxwell *The Scientific Papers of Peter Guthrie Tait* (London: Cambridge University Press)

Talbot, H F 1834 Experiments on light *Phil. Mag.* Ser. 3 **5**

Thompson, Benjamin 1774 An account of a method of measuring the comparative intensities of the light emitted by luminous bodies *Phil. Trans. R. Soc.*

Turbeville, Dawbeney 1684 Two letters from the great and experienced oculist Dr Turbeville of Salisbury to Mr William Musgrave, SPS of Oxon, containing several remarkable cases in physik, relating chiefly to the eyes *Phil. Trans. R. Soc.*

Tyndall, W H 1853 *The Athenaeum* Letter dated January 29 1853 on coloured signals on railroads

Waller, Richard 1685–6 A catalogue of simple and mixed colours *Phil. Trans. R. Soc.*

Wartmann, Èlie 1843 Memoire on Daltonism *Mem. Soc. Phys. Hist. Nat., Genève* **10** (English translation 1846 *Taylor's Scientific Memoirs* **4**)

Werner, Abraham 1774 *On the External Characters of Minerals* (English translation by Albert V Carozzi 1962 (Urbana: University of Illinois Press))

Wilson, George 1853 *The Athenaeum* Letter dated April 2 1853 in reply to

Tyndall's on railroad signals and colour blind drivers and signalmen
—— 1855 *Researches on Colour Blindness* (Edinburgh)
Wollaston, William Hyde 1802 A method of examining refractive and dispersive powers by prismatic reflection *Phil. Trans. R. Soc.*
Wünch, Chrétian-Ernst 1792 *Versuche und Beobachtungen über die Farben des Lichtes* (Leipzig) (French excerpt in 1807 *Ann. Chem., Paris* **64**)
Young, Thomas 1802 An account of some cases of the production of colours not hitherto described *Phil. Trans. R. Soc.*
—— 1802 *A Syllabus of a Course of Lectures on Natural and Experimental Philosophy* (London)
—— 1802 On the theory of light and colours *Phil. Trans. R. Soc.* **92**
—— 1804 Experiments and calculations relative to physical optics *Phil. Trans. R. Soc.*
—— 1817 Chromatics *Supplement to Encyclopaedia Britannica* (Edinburgh). Reprinted in 1855 *Miscellaneous Works of Thomas Young* ed G Peacock (London)
—— 1845 *A Course of Lectures on Natural Philosophy and the Mechanical Arts* ed P Kelland (London)

SECONDARY SOURCES

Birren, Faber 1965 *History of Colour in Painting* (New York: Reinhold)
—— 1969 *The Color Primer: A Basic Treatise on the Color System of Wilhelm Ostwald* (New York: van Nostrand Reinhold)
Bouma, P J 1936 The perception of color *Philips Tech. Rev.* **1**
Boyer, Carl B 1959 *The Rainbow: From Myth to Mathematics* (London: Thomas Yoseloff)
Campbell, Lewis and Garnett, William 1882 *The Life of James Clerk Maxwell* (London)
Caws, Peter 1965 *The Philosophy of Science: A Synthetic Approach* (Princeton: van Nostrand)
Forbes, E G 1970 Tobias Mayer (1723–62): a case of forgotten genius *Br. J. Hist. Sci.* **V**
—— 1970 Tobias Mayer's theory of colour mixing and its application to artistic reproductions *Ann. Sci.* **26**
Halbertsma, K T A 1949 *A History of the Theory of Colour* (Amsterdam: Swets and Zeitlinger)
Hanson, Norwood Russell 1961 *Patterns of Discovery* (London: Cambridge University Press)
Hecht, Selig 1930 Young's theory of colour vision *J. Opt. Soc. Am.* **20**
Henderson, S T 1969 *Daylight and Its Spectrum* (Bristol: Adam Hilger)
Henry, William Charles 1854 *Memoirs of the Life and Scientific Researches of John Dalton* (London)

Judd, D B 1961 Maxwell and modern colorimetry *J. Photog. Sci.* **9**

Judd, D B and Wyszecki, G 1963 *Color in Business Science and Industry* 2nd edn (New York: Wiley)

Kuhn, Thomas 1962 *The Structure of Scientific Revolutions* (Chicago: University of Chicago Press)

MacAdam, David L 1970 *Sources of Colour Science* (Cambridge, Ma: MIT Press)

Mach, Ernst 1926 *The Principles of Physical Optics, an Historical and Philosophical Treatment* (New York: Dover)

Merz, John Theodore 1923 *A History of European Thought in the Nineteenth Century* (Edinburgh and London: William Blackwood and Sons)

Millington, E C 1941 History of the Young–Helmholtz theory of vision *Ann. Sci.* **V**

Parsons, J Herbert 1915 *An Introduction to the Study of Colour Vision* (London: Cambridge University Press)

Peacock, George 1855 *Life of Thomas Young* (London)

Ronchi, Vasco 1970 *The Nature of Light* (English translation by V Barscas) (London: Heinemann)

Singer, H W 1903 Jakob Cristoffel LeBlon and his three-colour prints *The Studio* June

Walls, Gordon L 1943 The evolution of colour vision *J. Appl. Phys.* **14**

—— 1956 The G Palmer story *J. Hist. Med. Allied. Sci.* **XI**

Whewell, William 1847 *History of the Inductive Sciences from the Earliest to the Present Time* 2nd edn (London)

Williams, L Pearce 1962 The physical sciences in the first half of the nineteenth century: problems and sources *Hist. Sci.* **1**

Wolf, A 1962 *A History of Science, Technology and Philosophy in the Eighteenth Century* (London: George Allen and Unwin)

Wood, Alexander 1954 *Thomas Young, Natural Philosopher 1773–1829* (London: Cambridge University Press)

Wright W D 1946 *Researches on Normal and Defective Colour Vision* (London: Henry Kimpton)

—— 1967 *The Rays are Not Coloured* (Bristol: Adam Hilger)

—— 1968 The unsolved problem of Daltonism in *John Dalton and the Progress of Science* ed D S L Cardwell (Manchester: Manchester University Press)

Index